Appleton & Lange's Review of

EPIDEMIOLOGY & BIOSTATISTICS FOR THE USMLE

EDWARD J. HANRAHAN, MD
GANGADHAR MADUPU, MBBS, MS

APPLETON & LANGE
Norwalk, Connecticut

ISBN 0–8385–0244–X

Copyright © 1994 by Appleton & Lange
Paramount Publishing Business and Professional Group
Simon & Schuster Business and Professional Group

94 95 96 97 98 / 10 9 8 7 6 5 4 3 2 1

Prentice Hall International (UK) Limited, *London*
Prentice Hall of Australia Pty. Limited, *Sydney*
Prentice Hall Canada, Inc., *Toronto*
Prentice Hall Hispanoamericana, S.A., *Mexico*
Prentice Hall of India Private Limited, *New Delhi*
Prentice Hall of Japan, Inc., *Tokyo*
Simon & Schuster Asia Pte. Ltd., *Singapore*
Editora Prentice Hall do Brasil Ltda., *Rio de Janeiro*
Prentice Hall, *Englewood Cliffs, New Jersey*

Library of Congress Cataloging-in-Publication Data

Hanrahan, Edward J.
    Appleton & Lange's review of epidemiology & biostatistics for the USMLE / Edward J. Hanrahan, Gangadhar Madupu.
        p.   cm.
    Includes index.
    ISBN 0–8385–0244–X
    1. Epidemiology—Examinations, questions, etc.   2. Biometry—Examinations, questions, etc.   3. Statistics—Examinations, questions, etc.   4. Medical statistics—Examinations, questions, etc.   I. Madupu, Gangadhar.   II. Title.   III. Title: Appleton and Lange's review of epidemiology & biostatistics for the USMLE.   IV. Title: Epidemiology & biostatistics for the USMLE.
    [DNLM: 1. Epidemiology—examination questions.   2. Statistics—examination questions.   WA 18 H248a 1994]
    RA652.7.H36   1994
    614.4'076—dc20
    DNLM/DLC   94–7416
                                                     CIP

Acquisitions Editor: Jamie L. Mount
Production Editor: Sondra Greenfield
Designer: TopDesk Publishers' Group

PRINTED IN THE UNITED STATES OF AMERICA

ISBN 0-8385-0244-X

9 780838 502440

*This book is dedicated to all medical
students, residents, and fellows.*

Contents

Preface

This book has been written primarily for medical students and international medical graduates preparing for the United States Medical Licensing Examinations. Its aim is to present the core material that the medical student needs to successfully complete the epidemiology and biostatistics sections of those examinations. This book is not designed to be a comprehensive text but, rather, an outline of the essentials that can be read and reviewed in a limited amount of time.

The authors feel that the essentials of epidemiology and biostatistics can be mastered very quickly with the use of this review, and they suggest that it be read from cover to cover.

Gangadhar Madupu, MBBS, MS

Introduction

In the contemporary arena of medical certification, board examinations administered by the National Board of Medical Examiners (NBME) have become increasingly difficult. Medical students and international medical graduates are constantly on the lookout for a concise book in every subject, including epidemiology and biostatistics, to facilitate preparation for these examinations. The right kind of review must be broad in scope to permit attention to the important aspects of each discipline, yet concise enough to facilitate learning within time constraints. The United States Medical Licensing Examinations (USMLE) Step 1, Step 2, and Step 3 have been designed to measure comprehensive knowledge and the ability to apply that knowledge clinically. The NBME recommends that the most appropriate preparation is a general and thorough review of the basic and clinical sciences.

Most of the books written on the topics of epidemiology and biostatistics exceed 300 pages in length. They lack the brevity that time constraints necessitate in the burgeoning curriculum of the medical examinee. Because questions pertaining to epidemiology and biostatistics appear in all three steps of the USMLE, our endeavor is to introduce a short, concise review book that (1) clearly gets to the point on the essential material, (2) presents core material clearly and succinctly in a medically relevant and clinically understandable format, and (3) most important, tests the reader with board types of questions both within the topic discussions as well as in a separate timed examination.

The Editors and the Publisher

How to Use This Book

This book assists your preparation for the USMLE by providing an outline of study designs, statistical methods, and their clinical significance in contemporary medical practice. It illustrates how studies may be compared (ANOVA, chi-square, t tests, correlation, and linear regression analysis, and so on); how relative and attributable risks and odds ratios characterize independent variables as they relate to outcome (dependent variables); how a particular treatment regimen can be shown to be superior to another by using standard statistical methodologies; how primary prevention strategies reduce the incidence of disease, as well as the morbidity and mortality associated with secondary and tertiary prevention; and finally, how to view a graph, table, or chart to determine exactly what it conveys.

All these concepts are tested topics on Step 1 and Step 2 of the USMLE and are presented here for your review in a rapid reading format. The fundamental goal of this book is to deliver a conceptualization that can be readily assimilated and to equip you with the ability to quickly answer questions on epidemiology and biostatistics.

For those of you with a strong proficiency in these two areas of medicine, go straight to Appendix A, review the formulas, and proceed directly to the timed examination at the end of the book. After completing the test, check your answers and review the appropriate sections for the questions that were answered incorrectly.

For a more comprehensive review, begin with the first chapter and review each topic before taking the practice examination. Once you complete the subject review, take the practice examination. Check your answers, and review the sections (a second time) that are appropriate for the incorrectly answered questions.

The authors wish to express their confidence that the time taken by USMLE candidates to master this study guide will be time wisely spent.

Edward J. Hanrahan, MD

Acknowledgments

I would like to express my indebtedness to **Elizabeth Storm, PhD, MS,** chairperson of the Department of Epidemiology and Biostatistics at Ross Universiy School of Medicine, without whose distinguished lectures my interest in epidemiologic and biostatistical methodologies in medicine might never have materialized. To my closest friend, **Bill Beach,** I'll be eternally grateful for all his PC expertise and personal sacrifices that made this book a reality. And although it is impossible ever to adequately thank my family for all the support and encouragement that they have given me throughout the years, **Mom** . . . I love you.

Edward J. Hanrahan, MD

I am deeply indebted to the faculty of the Department of Social and Preventative Medicine (Epidemiology) of the State University of New York at Buffalo. Special thanks to **Tim E. Byers, MD, MPH,** a distinguished cancer epidemiologist, and **Maria Zielezny, PhD,** biostatistician, an outstanding teacher whose departmental lectures for master's-level graduate studies at the State University of New York at Buffalo served as the knowledge base for this book. I am also thankful to my family members, especially my brother Murali for his financial assistance during my years of schooling at Buffalo. Finally, I would like to thank all the **students** to whom I have taught epidemiology and biostatistics for providing me with the necessary experience to make this book possible.

Gangadhar Madupu, MBBS, MS

ESSENTIAL CONCEPTS

- Incidence
- Prevalence
- Case Fatality Rate
- Infant Mortality Rate
- Neonatal Mortality Rate
- Perinatal Mortality Rate
- Maternal Mortality Rate
- Age-Specific Mortality Rate
- Proportionate Mortality Rate
- Reliability
- Validity
- Sensitivity
- Specificity
- Positive Predictive Value
- Negative Predictive Value
- Prospective or Cohort Studies
- Retrospective or Case-Control Studies
- Prevalence or Cross-Sectional Studies
- Experimental Studies or Clinical Trials
- Relative Risk and Odds Ratio
- The Normal (Gaussian) Distribution
- Mean, Median, Mode, and Range
- Tests (t test, F test, and Chi-square)
- Correlation Coefficient and Regression Analysis
- Hypothesis Testing

CHAPTER 1

Epidemiological Terminology

1.1 Epidemiology

Epidemiology refers to the investigation of factors that determine the frequency and distribution of disease or other health-related conditions within a defined human population during a specified period.

1.2 Epidemics

An **epidemic** is an increase in the incidence of diseases, conditions, or other health-related events in a defined human population that is clearly in excess of that which was expected during a specified period. Although the presence of the disease or event is typically occasional in a particular community, its epidemicity is always relative to its usual frequency in terms of time, place, and population. Outbreaks that affect large numbers of animals are referred to as *epizootic*.

1.3 Endemics

Endemics are diseases, conditions, or health-related behaviors that are *constantly present* in a human population. An endemic may be referred to as the usual prevalence of an event or occurrence in a defined community. Endemics that occasionally or seasonally become epidemic are referred to as *endemoepidemics*, and those that are present at a high rate of incidence and affect all ages equally are called *hyperendemics*.

1.4 Pandemics

Pandemics are *widespread epidemics* that achieve large geographic proportions.

Example

During the influenza pandemic of the 19th century, millions of people across the continents were affected. Today, **AIDS** is considered to be the most alarming pandemic of the century.

1.5   Primary Prevention

Primary prevention is a reduction in the incidence of disease through immunization, sanitation, education, or other means of eliminating pathogenic contamination in the human environment.

Examples

Preschool diphtheria/pertussis/tetanus (DPT) immunization of children, polio and influenza vaccinations, and so on.

Q   1. All the following activities meet the criteria for primary prevention except

   (A) tetanus vaccination for teenagers
   (B) measles/mumps/rubella (MMR) vaccination for first-time-pregnant mothers
   (C) fluorification of water
   (D) pap smear screening
   (E) sex education

1.6   Secondary Prevention

Secondary prevention is the early detection and treatment of diseases.

Example

Pap smears for the early detection of cervical cancer and surgical intervention if necessary.

1.7   Tertiary Prevention

Tertiary prevention is the reduction of the complications of diseases and the improvement in the patient's level of function through palliative treatment and rehabilitation therapy.

Example

Shortly after the occurrence of hemiplegia secondary to stroke, bed care and physiotherapy are employed to prevent the development of decubitus ulcers and flexion contractures.

1.8   Spectrum of an Infectious Disease

The sequence of events beginning with the exposure of a susceptible individual to an pathogenic agent and ending with a patient's recovery or death is illustrated in the following diagram.

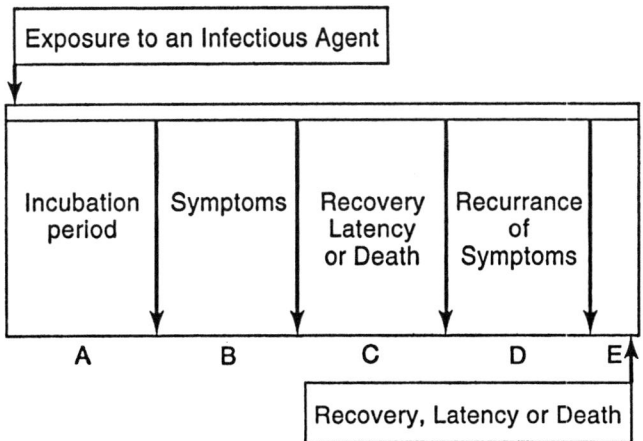

Fig. 1.1 Spectrum of Infectious Disease.

1. Incubation Period (*A* to *B*): The time interval between the invasion by an organism and the development of symptoms is referred to as the incubation period.

Example

After exposure to the measles virus and before the onset of symptoms, a child is considered to be highly contagious due to continued attendance at school and social encounters with other children.

2. Latent Period (*C* to *D*): This period is the interval of subclinical infection during which the previously active infectious agent becomes dormant in the host.

Example

Subsequent to the appearance of genital lesions induced by a herpes simplex Type II infection, patients often experience periods of remission, after which reactivation of the virus elicits the reappearance of lesions. This interval of remission is referred to as the latent period of the virus.

C H A P T E R   2

Classifications of Epidemics

Two fundamental assumptions of epidemiology are that disease neither occurs by chance nor is it distributed randomly in a population. Disease occurs at specific times in specific environmental locations and affects particular populations for very specific reasons. Painstaking explicitness being the key criterion in any epidemiologic investigation, when the outbreak of an epidemic is suspected, great care must be exercised both in the accumulation and the description of information regarding etiological agents, hosts, locations, chronologies, environmental factors, reservoirs, vectors, modes of transmission, number of cases, and so on. Only after a definitive diagnosis is confirmed and all of the above-mentioned factors relating to time, place, and person are described at length can the existence and identification of an epidemic be established and treatment and preventative measures be instituted.

Epidemics are classified as follows:

Common Source Epidemics
Point Source Epidemics
Propagative (Progressive) Epidemics

2.1   Common Source Epidemic

In **common source epidemics,** all susceptible individuals are exposed to a specific infectious pathogen or noxious agent (chemicals, pollution, heat, etc.) originating from a usual, conventional, or customary source—that is, common exposure. The mode of transmission is termed **indirect** because of the evidence that the pathogen or agent is **vehicleborne** (contacted through fomites, food, water, air, etc.) and because secondary host-to-host (*direct*) transmission is rare. By examining the shape of the epidemic curve, it can be seen that the fewest number of cases become apparent after a minimum incubation period and that the largest number peaks at the end of the usual incubation period (from the point of exposure to the midline of the curve), creating the typical unimodal shape of the common source epidemic curve. Although common source epidemics may involve only one incubation period of an organism, repeated or prolonged population exposure to the common source often involves infection by a pathogen over the

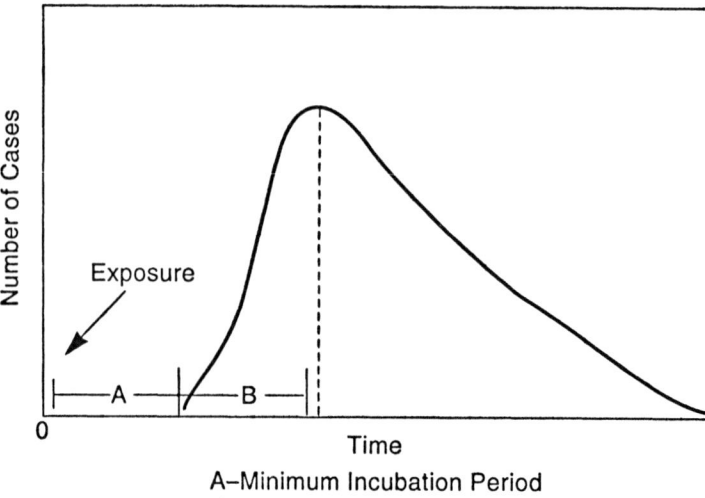

Fig. 2.1 Common Source Epidemic.

course of more than one incubation period, producing a wide peak or apex in the common source curve.

Examples

Infectious hepatitis (hepatitis A) transmitted by a food handler at a restaurant, or water contaminated with vibrio cholera at a local lake or estuary.

2.2   Point Source Epidemic

In a **point source epidemic,** all susceptible individuals are exposed to a specific pathogen at **one point in time.** Point source epidemics are essentially a *subcategory* of the **common source epidemic** in which common exposure to the offending pathogen or agent is both brief and simultaneous. As can be seen in the point source epidemic curve, there is a very explosive increase in

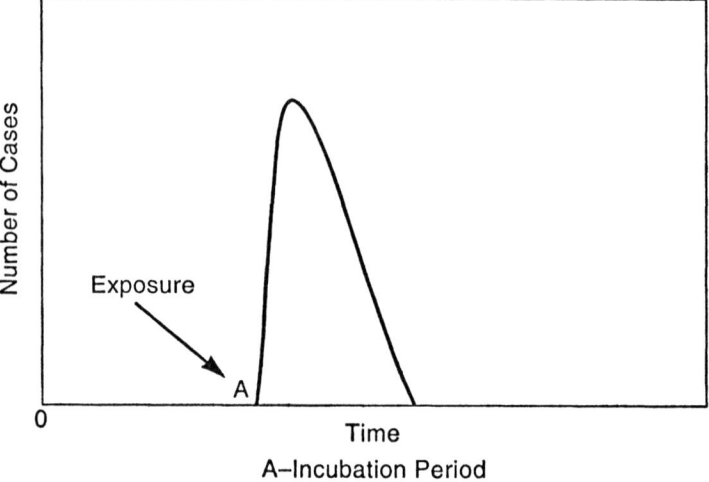

Fig. 2.2 Point Source Epidemic.

the number of cases over a short period of time after exposure to the noxious agent. Because the temporal aspect of the curve involves only one incubation period of the offending pathogen, the apex is much sharper and the decline in the number of cases is more rapid than in a typical common source curve. As with common source epidemics as a whole, secondary transmission is rarely seen in point source epidemics.

Example

After eating dinner at a wedding reception, many of the guests developed the acute onset of gastroenteritis 3 to 5 hours later as a result of food poisoning.

2.3 Propagative (Progressive) Epidemic

In **propagative (progressive, serial transmission) epidemics,** the pathogen is transmitted from **person to person.** The mode of transmission may be **direct,** in which the transfer of the infectious agent is through direct human contact (genital, anal, oral, and skin contact; directly aerosolized respiratory droplets; fungi, bacterial spores, and parasites), or **indirect,** in which the pathogen is either vectorborne (arthropods), or airborne (dried droplet residues and dust). In direct transmission, susceptibility to secondary infection is dependent upon the proximity of an individual to a contagious host, as well as to the portal of entry, pathogenicity, and dosage of the infecting organism. Individual immunogenicity and herd immunity also play very important roles in propagative (person-to-person) disease transmission. The curve for a propagative epidemic shows an initial rise in the number of cases that is less explosive than in a *point source epidemic,* and successive generations of secondary infection produce a polymodal distribution conforming to several generations of incubation periods.

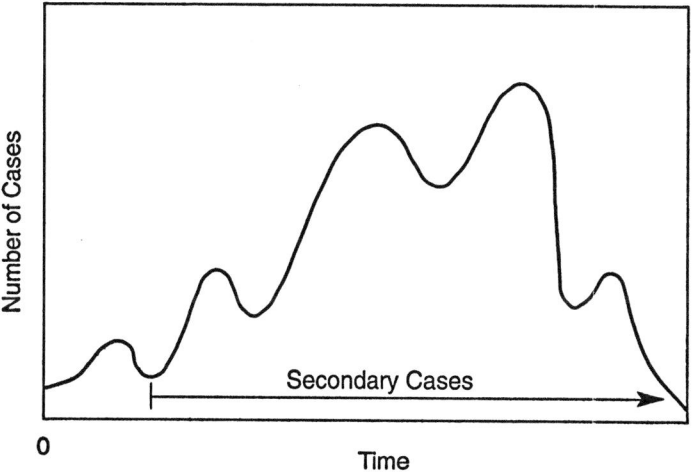

Fig. 2.3 Propagative (Progressive) Epidemic.

Examples

The classic example of propagative epidemics is measles transmission in school children. Among the many other examples are outbreaks of viral hepatitis, influenza, and sexually transmitted diseases (STDs). Examples of vectorborne (indirect) human-to-human transmission include malaria and yellow fever.

☐ NOTE

Although the shape of an epidemic curve may often help us to distinguish between **common source, point source, and propagative (progressive)** epidemics, several factors can make these distinctions very difficult. Repeated or prolonged exposure to a common source, recontamination of the common source, multiple incubation periods, long incubation periods, and the uncommon occurrence of secondary infection may all create a bi- or poly-modal distribution and obscure the typical unimodal shape of a *common (point) source* epidemic curve. Conversely, short incubation periods, as are seen in *propagative* epidemics like influenza, can introduce temporal relationships into the curve that are very similar to those of a *point source* epidemic. When such complications exist, a classification of the epidemic cannot be based solely upon the shape of the epidemic curve. Further investigation often involves a vectorial analysis of the geographic distribution of the outbreak to facilitate accurate classification.

Attack Rate

When a **point source epidemic** involves the occurrence of food poisoning, an identification of the specific food source (vehicle) that transmitted the infecting pathogen is made by comparing the attack rates between people who ate specific foods and those who did not. Similarly, when an outbreak of hepatitis A **(common source epidemic)** or measles **(propagative epidemic)** occurs, primary and secondary attack rates are computed for specific periods to determine the incidences of disease during these periods. When incidence rates are calculated for populations that are at risk for only a limited time (weekly, monthly, or for the duration of an epidemic lasting less than one year), they are commonly referred to as **attack rates.**

$$\text{Attack rate} = \frac{\text{Number of new cases during a specific period}}{\text{Number of people at risk during the same period}}$$

Example

During the month of November, a measles outbreak in an elementary school caused the absence of 100 of its 400 pupils. During the month of December, an additional 100 pupils were diagnosed with measles, as were 150 of the 200 brothers and sisters of the first 100 pupils.

Calculate: A) The attack rate of measles for students during the month of December.

B) The secondary attack rate for siblings.

A) $\text{Attack rate} = \dfrac{\text{Number of new cases during a specific period}}{\text{Number of people at risk during the same period}}$

$$= \dfrac{100}{400 - 100} = 33\%$$

B) Secondary attack rate $= \dfrac{150}{200} = 75\%$

□ NOTE

Always remember to reduce the number of people at risk by the number of people who are no longer at risk during the specified period.

Q 2. Shortly after a dormitory barbeque, medical students at Duke University came back to their rooms and most of them (62 out of 74 students) experienced acute vomiting and diarrhea. This epidemic may be classified as

(A) point source
(B) propagative
(C) common source
(D) serial
(E) direct

Q 3. Ten days after a measles outbreak in a small town, several elementary school children became symptomatic. Subsequently, additional cases were found among friends and families of the infected students. This epidemic may be classified as

(A) point source
(B) propagative
(C) common source
(D) indirect
(E) vectorborne

Q 4. After returning home from a family planning clinic, Dr. Cunningham noticed a slight itching between his fingers. Within 2 days his wife had similar itching, as did his son one day later. This epidemic (scabies) may be classified as

(A) point source
(B) indirect transmission
(C) common source
(D) serial transmission
(E) vehicleborne transmission

The 2-by-2 Table and Its Concepts

3.1 The 2-by-2 Table

A table that consists of two columns (vertical) that represent the presence or absence of a *disease* and two rows (horizontal) that represent a positive or negative *test result* is called a 2-by-2 table. These tables may also be employed in *risk factor studies* (page 35), where the rows represent the presence or absence of a risk factor, and in *hypothesis testing* (page 21), where the rows represent the acceptance or rejection of the null hypothesis.

The following 2-by-2 table is a tool for the evaluation of *standardized screening tests,* which predict the presence or absence of disease:

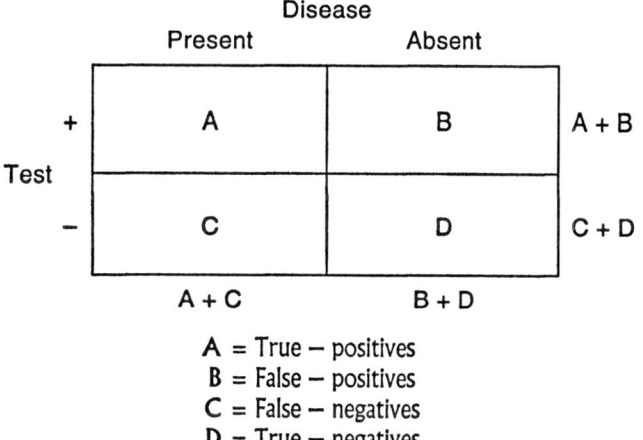

A = True − positives
B = False − positives
C = False − negatives
D = True − negatives

3.2 Reliability and Precision

Reliability is the *dependability* of a test result. **Precision** is the extent to which *random variability* is *absent* from a test result. The **reliability** of a test is directly dependent upon its **precision,** because random variation in a test result precludes the test's ability to be *consistent* and *reproducible,* and therefore reduces its *dependability.* The **precision** (*consistency* and *reproducibility*) of a test are the characteristics that determine the degree to which it may be considered **reliable** (*durable* and *dependable*). *Diminished precision* in a test is the result of what is commonly referred to as **random error.**

3.3 Validity and Accuracy

Validity is the extent to which a test *measures what it was designed to measure*. **Accuracy** is the ability of a test to produce a *true value for the measurements* and *true classifications* for the samples under study. Screening tests should be able to identify the *presence* or *absence* of a specific disease **(validity).** They can do so only with *true measurements* of the markers of that particular disease and *true classifications* of *sample populations* **(accuracy).** Diminished accuracy in a test or study is most often the result of a *nonrandom (systemic)* type of experimental error called **bias.**

Validity has two components:

A. **Sensitivity** is defined as the ability to *correctly identify individuals who have a specific disease or condition.*

B. **Specificity** is defined as the ability to *correctly identify individuals who do not have a specific disease or condition.*

Although the *validity* of a diagnostic test can be determined by *sensitivity* and *specificity,* three other measures are important:

C. The **false-positive rate** is defined as the proportion of false positives (*B*) among nondiseased (*B* + *D*) subjects.

D. The **false-negative rate** is defined as the proportion of false negatives (*C*) among diseased (*A* + *C*) subjects.

E. The **accuracy of a test** is defined as the proportion of true results [ie, true positives (*A*) + true negatives (*D*)] among all test results.

With this background knowledge, let's master the first five components of the **screening test:** sensitivity, specificity, false-positive rate, false-negative rate, and accuracy of test.

		Disease		
		Present	Absent	
Test	+	A	B	A + B
	−	C	D	C + D
		A + C	B + D	

3.4 Sensitivity $A/(A + C)$

3.5 Specificity $D/(B + D)$

3.6 False-Positive Rate $B/(B + D)$

3.7   False-Negative Rate $C/(A + C)$

3.8   Accuracy $(A + D)/(A + B + C + D)$

3.9   ROC (Receiver Operating Characteristic) Curve

Receiver operating characteristic (ROC) curves are graphic illustrations of the *differences between the sensitivity and specificity of two diagnostic tests* that are viewed for the purpose of comparing their validity and effectiveness. By plotting the *true-positive rates (sensitivity)* on the y axis and their corresponding *false-positive rates (1-specificity)* on the x axis, we are able to select the more accurate of the two tests by selecting the curve that lies closer to the *upper left-hand corner* of the graph—that is, where the *true-positive rate* (sensitivity) = 1, and the *false-positive rate* (1-specificity) = 0.

Example

Comparing the *sensitivity* and *specificity* of two diagnostic tests for syphilis, a researcher plotted the following ROC curves for the **VDRL** (Venereal Disease Research Laboratory) and **FTA-ABS** (Fluorescent Treponemal Antibody Absorption) tests. As you can see, the FTA-ABS test curve lies closer to the *upper left-hand corner* of the graph where the *true-positive rate (sensitivity)* = 1 and the *false-positive rate (1-specificity)* = 0. It was therefore found to be both more sensitive and more specific than the VDRL test for

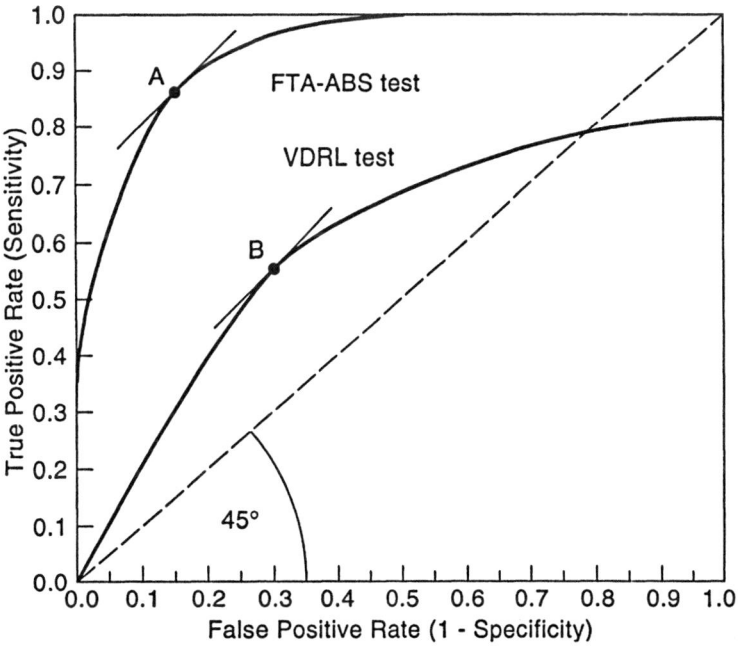

Fig. 3.1 ROC (Receiver Operating Characteristic) Curve.

the detection of syphilis. Points A and B of the FTA-ABS and VDRL tests respectively represent the best **cutoff point** for each test (discussed later) and can be determined by locating the point where the tangent to the line of the curve is equal to 45 degrees. The *accuracy* of the test is determined by measuring the area between the curve and a 45-degree line drawn from the origin of the graph.

3.10   Effects of the Cutoff Point on Sensitivity and Specificity

The **cutoff point** of a screening test has an effect on the *sensitivity* and *specificity* of the test and can be demonstrated by the following example: Let's define an appropriate *cutoff point* for a diagnostic screening test (fasting blood glucose) to detect the presence or absence of diabetes mellitus. If we set the cutoff point of the diagnostic test too high (point **B** in the following figure), the test will miss the detection of mild hyperglygemia. Conversely, setting the cutoff point of the screening test too low (point **A** in the following figure) will cause the test to detect some of the normal subjects (absence of diabetes) as being hyperglycemic. A summary of these effects is given in the following figure.

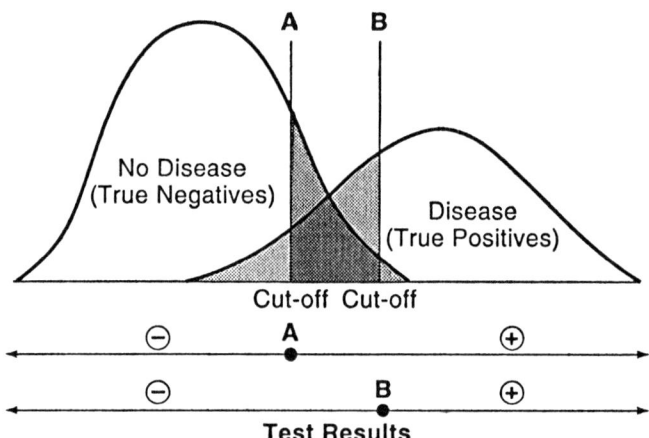

Fig. 3.2 Cut-off Points in a Screening Test.

Cutoff Point A

- **Greater sensitivity**
- **Greater false-positive rate**
- **Lower specificity**

Cut off Point B

- **Lower sensitivity**
- **Greater false-negative rate**
- **Greater specificity**

3.11   The Predictive Value of a Screening Test

The **predictive value** of a screening test measures the *true presence* or *absence of a disease*. Predictive value has two components:

A. The **positive predictive value** is the proportion of true positives among all positives.

B. The **negative predictive value** is the proportion of true negatives among all negatives.

With this background knowledge, let us master two more components of the screening test: positive predictive value and negative predictive value.

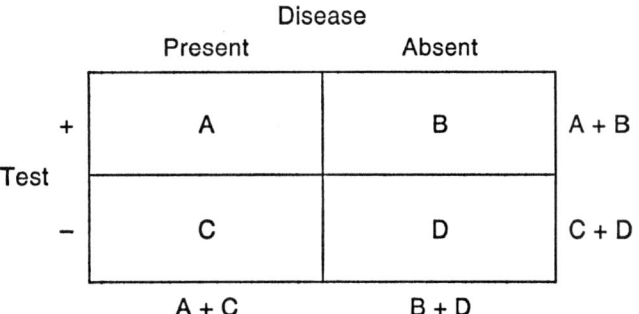

3.12   Positive Predictive Value = $A / (A + B)$

3.13   Negative Predictive Value = $D / (C + D)$

3.14   The Relationship Between Prevalence and the Predictive Value of a Test

In general, the usefulness of a diagnostic screening test is affected by the **prevalence** of a disease: If the *prevalence* of a disease is *low,* the *positive predictive value* is likely to be *low;* if the *prevalence* of a disease is *high,* the *positive predictive value* is likely to be *high.*

Example

The enzyme-linked immunosorbent assay (**ELISA**) is being employed in the screening of routine blood donors for the detection of **HIV**-positive antibodies. By using the ELISA, the prevalence of HIV antibodies among those screened was found to be less than 1%. This *low prevalence* will yield a *low positive predictive value* for a screening test. Conversely, the **VDRL** is being employed in the routine screening of communities for the detection of **STDs** (sexually transmitted diseases). By using the VDRL test, the prevalence of STDs among those screened was found to be high. This *high prevalence* will yield a *high positive predictive value* for the screening test.

2-BY-2 TERMINOLOGY

		Disease		
		Present	Absent	
Test	+	A	B	A + B
	−	C	D	C + D
		A + C	B + D	

Sensitivity	$= A/(A + C)$	$= \dfrac{\text{true positives}}{\text{true positives} + \text{false negatives}}$
Specificity	$= D/(B + D)$	$= \dfrac{\text{true negatives}}{\text{false positives} + \text{true negatives}}$
False-positive rate	$= B/(B + D)$	$= \dfrac{\text{false positives}}{\text{false positives} + \text{true negatives}}$
False-negative rate	$= C/(A + C)$	$= \dfrac{\text{false negatives}}{\text{true positives} + \text{false negatives}}$
Positive predictive value	$= A/(A + B)$	$= \dfrac{\text{true positives}}{\text{true positives} + \text{false positives}}$
Negative predictive value	$= D/(D + C)$	$= \dfrac{\text{true negatives}}{\text{true negatives} + \text{false negatives}}$
Accuracy	$= \dfrac{A + D}{(A + B + C + D)}$	$\dfrac{\text{true positives} + \text{true negatives}}{\text{all positives} + \text{all negatives}}$

Example

An investigator evaluated 100 patients suffering from major depression as confirmed by the attending psychiatrist. The results were as follows:

		CLINICAL DEPRESSION		
		PRESENT	ABSENT	
Test	+	12	18	(12 + 18)
	−	28	42	(28 + 42)
		(12 + 28)	(18 + 42)	N = 100

The following ratios were calculated by the investigator:

(A) $12 / (12 + 28)$
(B) $18 / (18 + 42)$
(C) $42 / (18 + 42)$
(D) $12 / (12 + 18)$
(E) $42 / (28 + 42)$
(F) $28 / (12 + 28)$
(G) $(12 + 42) / (12 + 18 + 28 + 42)$

Match the ratios with the following numbered items:

Q   **5.** Sensitivity

Q   **6.** Specificity

Q   **7.** Positive predictive value

Q   **8.** Negative predictive value

Q   **9.** False-positive rate

Q   **10.** False-negative rate

Q   **11.** Accuracy of a test

Q   **12.** The extent to which a test measures what it was originally designed to measure is described as

(A) sensitivity
(B) specificity
(C) validity
(D) reliability
(E) true-positive value

Match the following lettered options with the following numbered items:

(A) validity
(B) systemic error
(C) reliability
(D) random error
(E) precision

Q   **13.** Depends upon precision

Q   **14.** Dependability of a test

Q   **15.** Depends upon accuracy

Q 16. Consistency of a test

Q 17. Limited precision in a test

Q 18. Limited accuracy in a test

For questions 19 to 22, refer to the following table:

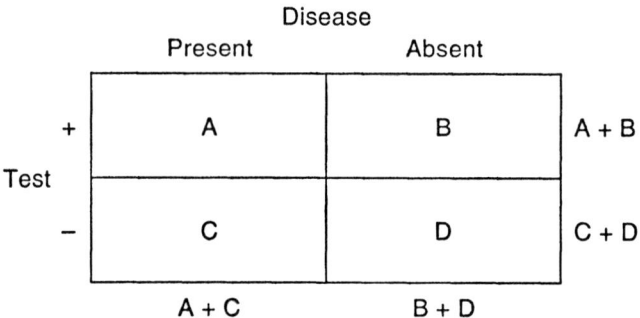

Q 19. False negatives are represented by

(A) A (B) B (C) C (D) D

Q 20. False positives are represented by

(A) A (B) B (C) C (D) D

Q 21. True negatives are represented by

(A) A (B) B (C) C (D) D

Q 22. True-positives are represented by

(A) A (B) B (C) C (D) D

E x a m p l e

The following two hypothetical distribution curves represent the effects of altering the **cutoff point** for serum cholesterol levels in a test for the classification of individuals who are at high risk for developing coronary artery disease:

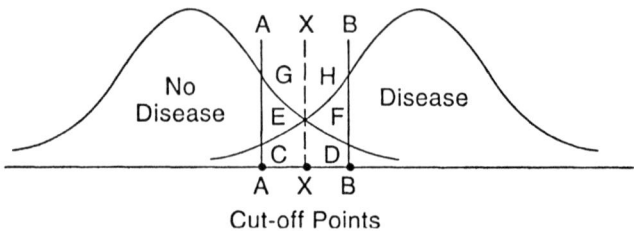

Fig. 3.3 Diagnostic Screening Test.

Match the following numbered items with the designated letters of the figure given in the preceding example:

Q   23. Cutoff point set too low

Q   24. Cutoff point set too high

Q   25. Cutoff point of greater sensitivity

Q   26. Cutoff point of lesser sensitivity

Q   27. Cutoff point of greater specificity

Q   28. Cutoff point of lesser specificity

Q   29. Cutoff point of greater false positive rate

Q   30. Cutoff point of greater false negative rate

Q   31. True positives for cutoff point X

Q   32. True positives and false positives for cutoff point X

Q   33. True negatives for cutoff point X

Q   34. True negatives and false negatives for cutoff point X

Q   35. False positives for cutoff point A

Q   36. False negatives for cutoff point B

Hypothesis Testing

Hypothesis testing is of paramount importance in medical research because it permits the researcher to make generalizations about a population based upon probabilities obtained from sample study results.

The aim of the researcher is to demonstrate that the observed findings obtained from a study were statistically significant. Hypothesis testing confirms (or refutes) the assertion that the observed findings did not occur by chance alone but by a true association between the dependent and independent variables.

For example, a researcher studied the relationship between **smoking** and the development of **lung cancer.**

		Lung Cancer	
		Present	Absent
Smoking	+	A	B
	−	C	D

4.1   Null Hypothesis (H₀)

In this study, the **null hypothesis (H$_0$)** states that **there is no difference** between smokers and nonsmokers with respect to the risk of developing lung cancer. *The observed difference, if any, is* **by chance alone.** The hypothesis that the researcher wants to test in the study is that smokers are at a higher risk than nonsmokers of developing lung cancer.

4.2   Alternate Hypothesis (H$_A$)

The **alternate hypothesis (H$_A$)** states that **there is a difference** between smokers and nonsmokers with respect to the risk of developing lung cancer and that *the observed difference is not by chance.* If the findings of the study are statistically significant and the null hypothesis has not been shown to be true, H_0 can then be rejected and the alternate hypothesis (H_A) accepted.

4.3   Types of Errors

	Truth	
	H_0 True	H_0 False
Accept H_0	Correct	Type II error
Reject H_0	Type I error	Correct

Decision

DIFFERENCES
H_0 True = statistically insignificant
H_0 False = statistically significant
Accept H_0 = statistically insignificant
Reject H_0 = statistically significant

Type I (Alpha) Error: If the H_0 is true in reality and the observed finding of a study is statistically insignificant, it is a correct decision to accept the null hypothesis. On the other hand, if H_0 is true and the observed finding of a study is statistically significant, the decision to reject H_0 is incorrect and an error has been made. This is called a Type I, or alpha, error. Therefore, a Type I error is **rejecting the null hypothesis (H_0) when it is true.**

Type II (Beta) Error: If in reality the H_0 is false and the observed finding of a study is statistically significant, it is a correct decision to reject the null hypothesis. On the other hand, if the H_0 is false and the observed finding of a study is statistically insignificant, the decision to accept H_0 is incorrect and an error has been made. This is called a Type II, or beta, error. Therefore, a Type II error is **accepting the null hypothesis (H_0) when it is false.**

☐ NOTE

The **power of a test** (probability that a test detects differences that *actually exist*) can be determined by using the formula **1 – beta ($1 - \beta$).** 80% is usually acceptable.

4.4   Level of Significance (*p* Value)

The **level of significance** in a study is the probability (*p*) that represents *the lowest significance level at which the null hypothesis (H_0) can be rejected. Most researchers use* **p < 0.05** (less than

5%) to reject H_0, which is fairly arbitrary but universally accepted. Conversely, when the probability *exceeds* 0.05 ($p > 0.05$), the null hypothesis is accepted and the alternate hypothesis rejected.

☐ NOTE

Although **statistical significance** can be a *true association,* it can also be *artifactual* due to a **confounding factor** (discussed in Chapter 6). Therefore, statistical significance does not necessarily prove either a *causal relationship* or *clinical significance.*

Q 37. In a medical journal report, the observed mortality of smokers and nonsmokers for laryngeal squamous cell carcinoma was reported to be significant at $p < 0.05$. Such a statement means that

(A) the investigator is rejecting the null hypothesis even though the results could have occurred purely by chance a maximum of 5 times out of 100

(B) there is a difference between the mortality rates of smokers and nonsmokers 5% of the time

(C) the null hypothesis claims that there is a difference between the mortality rates of smokers and non-smokers

(D) a causal relationship between smoking and mortality may be established through this study

(E) there are insufficient data, as the total number of smokers and nonsmokers were not given

Q 38. What is the claim of the null hypothesis (H_0) in this study?

(A) The null hypothesis (H_0) claims that there is a difference between the mortality rates of smokers and nonsmokers.

(B) The null hypothesis (H_0) claims that there is no difference between the mortality rates of smokers and nonsmokers.

(C) The null hypothesis (H_0) claims that there is a difference between the mortality rates of smokers and nonsmokers 5% of the time.

(D) The null hypothesis (H_0) claims that there is no difference between the mortality rates of smokers and nonsmokers 5% of the time.

(E) The null hypothesis (H_0) claims that there is no difference between the mortality rates of smokers and nonsmokers 95% of the time.

Q 39. What is the claim of the alternate hypothesis (H_A) in this study?

(A) The alternate hypothesis (H_A) claims that there is no difference in mortality between smokers and non-smokers.

(B) The alternate hypothesis (H_A) claims that there is a difference in mortality between smokers and non-smokers 5% of the time.

(C) The alternate hypothesis (H_A) claims that there is a difference in mortality between smokers and non-smokers 95% of the time.

(D) The alternate hypothesis (H_A) claims that there is no difference in mortality between smokers and non-smokers 5% of the time.

(E) The alternate hypothesis (H_A) claims that there is no difference between the mortality rates of smokers and nonsmokers 95% of the time.

4.5 Probability Theory

Multiplication Rule

To determine the probability (*P*) of the combined occurrence of two *independent* events (*A* and *B*), the multiplication rule is used. Events *A* and *B* are said to be *independent* if the occurrence of one event has no effect upon the occurrence of the other. The probability that both independent events will actually occur is the product of the probabilities of each event, and is calculated by the formula

$$P(A \text{ and } B) = P(A) \times P(B)$$

Example

If tuberculous meningitis had a case fatality rate of 20%, the probability that this disease would be fatal in two randomly selected patients is found by applying the **multiplication rule,** because the events are independent:

$$
\begin{aligned}
P(A \text{ and } B) &= P(A) \times P(B) \\
&= 20\% \times 20\% \\
&= 0.04
\end{aligned}
$$

☐ NOTE

When both *A* and *B* cannot occur together, they are referred to as being *mutually exclusive.* When this is true, *P* (*A* and *B*) = 0.

Addition Rule

Mutually exclusive events: To determine the probability (*P*) that one of two *mutually exclusive* events occurs (*A* or *B*), the addition rule is used. Events *A* and *B* are said to be mutually exclusive when only one of the events may occur—that is, *P*(*A* and *B*) = 0. The probability that only one of two *mutually exclusive* events will actually occur is the sum of the probabilities of each event and is calculated by the formula

$$P(A \text{ or } B) = P(A) + P(B)$$

Example

In a normally distributed population, the probability that a subject's serum cholesterol level will be lower than one standard deviation below the mean or higher than two standard deviations above the mean is found by applying the **addition rule** because the events are mutually exclusive. Because 68% of the cholesterol levels are found within one standard deviation of the mean, 32% of the levels are located outside this area: 16% above, and 16% below. Similarly, since 95% of the values are found within two standard deviations of the mean, 5% are located outside this area: 2.5% above, and 2.5% below. Therefore,

$$
\begin{aligned}
P(A \text{ or } B) &= P(A) + P(B) \\
&= 16\% + 2.5\% \\
&= 18.5\%
\end{aligned}
$$

Mutually inclusive events: To determine the probability (P) that one or both events occur (A, B, or A and B)—that is, at least one event occurs—the addition rule is again used. The probability that at least one of two events occurs is the sum of the probabilities of each event minus the probability that both occur together and is calculated by the formula

$$P(A \text{ or } B \text{ or both}) = P(A) + P(B) - P(A \text{ and } B)$$

Example

If tuberculous meningitis had a case fatality rate of 20%, the probability that this disease would be fatal in at least one of the two randomly selected patients is:

$$
\begin{aligned}
P(A \text{ or } B \text{ or both}) &= P(A) + P(B) - P(A \text{ and } B) \\
&= (0.2 + 0.2) - (0.2 \times 0.2) \\
&= 0.4 - 0.04 \\
&= 0.36
\end{aligned}
$$

Compare this with the example given for the multiplication rule.

4.6 One-Tailed (One-Sided) Testing

In **one-tailed testing,** the researcher is concerned with differences in *only one direction* ($A > B$ or $A < B$) from the mean.

Example

In a dietary study of first-year medical students at Temple University, blood samples were taken to record the initial mean serum cholesterol level of the class. Then their diet was replaced with a high-fiber menu. Semi-annual blood samples were then taken for three years to measure the *decrease* in the mean serum cholesterol level of the class before they graduated. Here, the researcher is investigating the capacity for change in only one direction. This is called *one-tailed (one-sided) testing*.

☐ NOTE

The same would be true of any interest in *increased* values of a variable.

4.7   Two-Tailed (Two-Sided) Testing

In **two-tailed testing,** the researcher is concerned with the difference in *both directions* (A > B and A < B) from the mean.

Example

A new antidepressant drug was given to a group of patients suffering from major depression. After 6 months, the researcher wanted to determine how many patients showed a beneficial effect from the drug as opposed to a deleterious effect. Here, the researcher is interested in knowing both the improvements and the detriments (side effects) of the drug being tested. This is called *two-tailed (two-sided) testing*.

Measures of Morbidity and Mortality

A. Measures of Morbidity

5.1 Incidence Rate

$$\text{Incidence Rate} = \frac{\text{Number of new cases during a specific period}}{\text{Total midperiod population at risk}}$$

High incidence is most often a result of a higher than average risk of developing a particular disease.

Example

People with AIDS have a higher *incidence* of Kaposi's sarcoma and pneumocystis carinii pneumonia (PCP) than the general population.

Example

People who smoke cigarettes have a higher *incidence* of lung cancer when compared to nonsmokers.

Example

Among the inpatients of Morris County Hospital during the month of June, a total of 12 were admitted with a primary diagnosis of UTI (urinary tract infection). For the same month, the hospital had a total of 2400 patients. The *incidence rate* of UTI for the month of June per 1000 patients is

$$(12/2400) \times 1000 = 5 \text{ per 1000 patients}$$

5.2 Prevalence Rate

$$\text{Point prevalence} = \frac{\text{Number of existing cases at a specific point of time}}{\text{Total midpoint population at risk}}$$

$$\text{Period prevalence} = \frac{\text{Number of existing cases at a specific period of time}}{\text{Total midperiod population at risk}}$$

Prevalence usually refers to *point prevalence,* and high prevalence normally refers to diseases of a *chronic* nature—that is, diseases of long duration.

Example

Prevalence of diabetes is higher in people over 65 years of age.

Example

During the month of April, a total of 32 patients in Morris County Hospital were found to have diabetes mellitus, and the hospital had a total of 6400 patients. The *prevalence rate* of diabetes for the month of April per 1000 patients is

$$(32/6400) \times 1000 = 5 \text{ per 1000 patients}$$

5.3 Relationship Between Incidence and Prevalence

Prevalence = Incidence × Duration of disease
$$P = I \times DD$$

Q 40. The incidence of bronchogenic carcinoma in Putnam County was found to be three times higher for men than for women, but the prevalence rates were almost the same in both sexes. This can be explained by:

(A) The duration of the disease is shorter for women
(B) The incidence rate is higher for women
(C) The recovery rate from the disease is higher for women
(D) The duration of the disease is becoming shorter for men
(E) The duration of the disease is becoming longer for men

Annual incidence rate of disease = 20 per 10,000,000
Annual mortality rate of disease = 4 per 10,000,000
Point prevalence rate of disease = 80 per 10,000,000

Q 41. With the information given, what is the average duration of the disease?

(A) 1 year
(B) 2 years
(C) 4 years
(D) 5 years
(E) Insufficient data

The **incidence** and **prevalence rates** of pneumococcal pneumonia in children under five years of age in a defined community are illustrated in the following figure:

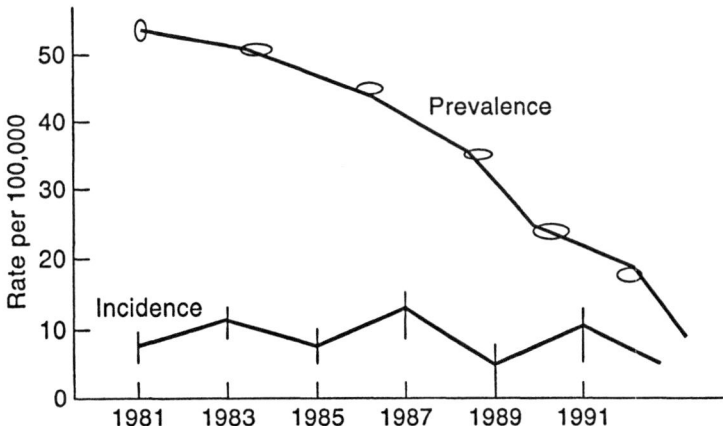

Fig. 5.1 Pneumococcal Pneumonia in Children Under Age Five.

Q 42. Possible conclusions that can be obtained from the data include:

(A) Primary prevention is successful due to vaccination
(B) The duration of the disease is increasing
(C) Incidence is decreasing
(D) Recovery from the disease is becoming rapid
(E) The rate of recovery remains the same

5.4 Person-Years

The term **person-years** is defined as the number of persons followed over a specified period of time in years. The period of observation may be less than one year but it is always expressed in terms of years. It is defined as:

Number of Persons × Number of Years Followed

E x a m p l e

In a cohort study, 600 persons were followed over the course of three years as follows:

100 people for 1 year
200 people for 2 years
300 people for 3 years

The number of person-years of observation in this study is calculated as follows:

100 persons × 1 year = 100 person-years
200 persons × 2 years = 400 person-years
300 persons × 3 years = 900 person-years
Total   = 1400 person-years

Q 43. The measure that is frequently used as a denominator to calculate the incidence rate is

(A) Number of new cases
(B) Number of existing cases
(C) Total population not at risk
(D) Person-years of observation
(E) Number of new and existing cases

B. Measures of Mortality

5.5 Case Fatality Rate

$$\frac{\text{Number of deaths due to a disease}}{\text{Number of people with the same disease}} \times 100$$

which is usually expressed as a percentage.

Example

In Bergen County, there were 500 HIV-positive people of whom five died within a year after their initial diagnosis. The *case fatality rate* is

$$5/500 \times 100 = 1\%$$

5.6 Infant Mortality Rate*

$$\frac{\text{Number of infant deaths less than one year of age}}{\text{Total number of live births during the same year}} \times 1000$$

Example

Denver, Colorado, with a population of 2.37 million, reported a total of 270 infant deaths and 30,000 live births in 1981. The *infant mortality rate* is

$$\frac{270}{30,000} \times 1000 = 9$$

or 9 infant deaths per 1000 live births per year.

5.7 Neonatal Mortality Rate**

$$\frac{\text{Number of neonatal deaths}}{\text{Total number of live births during the same year}} \times 1000$$

---

* The term **infant mortality** is completely exclusive of stillbirths.
** The *neonatal period* is defined as the interval of time between birth and 28 days postpartum.

Example

Memphis, Tennessee, with a population of 1.37 million, reported a total number of 150 neonatal deaths and 30,000 live births in 1982. The *neonatal mortality rate* is

$$\frac{150}{30,000} \times 1000 = 5$$

or 5 *neonatal deaths* per 1000 live births per year.

5.8   Perinatal Mortality Rate*

$$\frac{\text{Number of perinatal deaths}}{\text{Total number of live births during the same year}} \times 1000$$

Example

Cleveland, Ohio, with a population of 1.87 million, reported a total number of 280 perinatal deaths and 70,000 live births in 1983. The *perinatal mortality rate* is

$$\frac{280}{70,000} \times 1000 = 4$$

or 4 perinatal deaths per 1000 live births per year.

5.9   Maternal Mortality Rate**

$$\frac{\text{Number of deaths from puerperal causes***}}{\text{Total number of live births during the same year}} \times 100,000$$

Example

Los Angeles, California, with a population of 8.07 million, reported a total number of 42 maternal deaths attributed to puerperal causes, and 700,000 live births in 1984. The *maternal mortality rate* is

$$\frac{42}{700,000} \times 100,000 = 6$$

or 6 maternal deaths per 100,000 live births per year.

---

* The *perinatal period* is defined as the interval of time between 20 to 28 weeks gestation and 1 to 4 weeks postpartum.

** In the strictest sense, *maternal mortality rate* is a *ratio* of pregnancy-related deaths to live births, usually per 100,000.

*** The *puerperal period* is defined as the interval of time between birth and 4 to 6 weeks postpartum.

5.10   Annual Crude Mortality Rate

$$\frac{\text{All deaths during a calendar year}}{\text{Total midyear population}} \times 1000$$

Example

Atlanta, Georgia, with a population of 2 million, reported a total of 12,000 deaths from all causes in 1985. The *annual crude mortality rate* is

$$\frac{12,000}{2,000,000} \times 1000 = 6$$

or 6 deaths per 1000 population per year.

5.11   Age-Specific Mortality Rate

$$\frac{\text{Number of people who died in a particular age group}}{\substack{\text{Total midyear population of the same age group during} \\ \text{the same year}}} \times 1000$$

TABLE 5.1 PNEUMOCYSTIS CARINII PNEUMONIA
IN MIAMI, FLORIDA, 1989

AGE GROUP IN YEARS	CASES IN MIAMI	POPULATION OF MIAMI	DEATHS IN MIAMI	STATE POPULATION OF FLORIDA
17 to 27	100	50,000	50	1,000,000
28 to 47	300	150,000	750	2,500,000
> 48	350	250,000	1000	4,000,000
Total	800	450,000	1800	7,500,000

Example

The group that has the highest age-specific mortality rate from pneumocystis carinii pneumonia per 1000 population is found as follows:

17–27 (50/50,000)   × 1000 = 1.0 per 1000 population
28–47 (750/150,000) × 1000 = 5.0 per 1000 population
> 48 (1000/250,000) × 1000 = 4.0 per 1000 population

5.12   Cause-Specific Mortality Rate

$$\frac{\text{Number of deaths due a specific disease}}{\text{Total midyear population}} \times 1000$$

Example

There were 6000 deaths (1250 leukemia, 1750 stroke, 250 pneumonia, 250 Hodgkin's disease, and 2500 acute myocardial infarction) in Cook County, which had a population of 2.5 million in

1991. The cause-specific *mortality rate* for acute myocardial infarction is:

$$(2500/2,500,000) \times 1000 = 1 \text{ per 1000}$$

5.13 Proportionate Mortality Rate

$$\frac{\text{Total number of deaths due to a certain disease}}{\text{Total number of deaths from all causes}} \times 100$$

which is usually expressed as a percentage.

Example

There were 6000 deaths (1200 leukemia, 1750 stroke, 250 pneumonia, 250 Hodgkin's disease, and 2500 acute myocardial infarction) in Cook County, which had a population of 2.5 million in 1991. The *proportionate mortality rate* for leukemia is:

$$(1200/6000) \times 100 = 20\%$$

Refer to this table for the following examples and question:

TABLE 5.2 NUMBER OF CASES AND DEATHS
FROM HISTIOCYTOSIS X IN 1991

AGE GROUP IN YEARS	FOX COUNTY CASES	FOX COUNTY DEATHS	FOX COUNTY POPULATION
1–30	10	5	2,000
31–60	50	30	5,000
61+	30	20	1,000
Total	90	55	8,000

Example

The annual incidence rate for histiocytosis X in Fox County per 100,000 population is:

$$(90/8,000) \times 100,000 = 1125 \text{ per 100,000 population.}$$

Example

The *crude mortality rate* for histiocytosis X in Fox County per 100,000 population is

$$(55/8,000) \times 100,000 = 687.5 \text{ per 100,000 population}$$

Example

Age-specific mortality rate in people between 31 and 60 years of age for histiocytosis X in Fox County per 1000 population is

$$(20/5000) \times 1000 = 4 \text{ per 1000 population}$$

Example

The group that has the highest case fatality rate for histiocytosis X in Fox County is

$$
\begin{array}{rrcl}
0\text{--}30: & 5/10 & = & 50\% \\
31\text{--}60: & 30/50 & = & 60\% \\
61+: & 20/30 & = & 66\%
\end{array}
$$

CHAPTER 6

Epidemiological Study Designs and Measures of Risk

1. Prospective or Cohort Study
2. Retrospective or Case-Control Study
3. Cross-Sectional or Prevalence Study
4. Experimental Studies or Clinical Trials

6.1 Prospective or Cohort Study*

This study is also known as a **longitudinal** or **incidence** study design.

In a **prospective** or **cohort study,** a group of people are followed over a specified period to determine how many develop a specific disease or condition (**incidence**) after exposure to the risk factor or attribute under study. A **cohort** is a group of people who are either of the same age or share some other common characteristic(s).

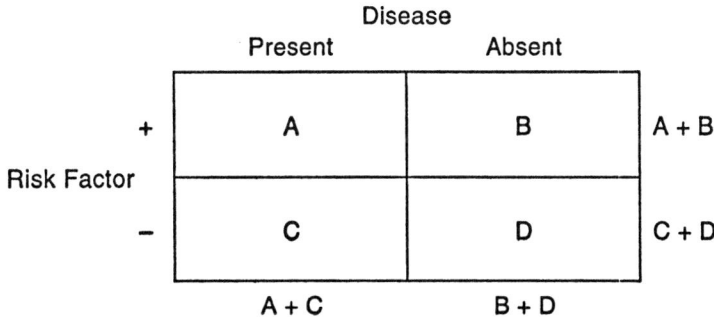

Prospective studies are usually *community-based, time-consuming, expensive, difficult to perform,* and *require a larger number of subjects* than **case-control studies** (discussed later). In a prospective study, the **incidence** and **relative risk** determina-

---

* This study attempts to answer the question, "What should happen in the future?"

tions can be made with accuracy and subjects are *less prone to selection bias* than in case-control studies. Prospective or cohort studies are usually done for relatively *common diseases*.

Q   44. Current theories about the possible relationship between high serum cholesterol levels and coronary artery disease were primarily established through the use of

(A) census taking
(B) case-control studies
(C) prospective population cohort studies
(D) patients with two or more myocardial infarctions
(E) experimental studies of laboratory animals

6.2   The Relative Risk in a Prospective Study*

In a **prospective study** of the relationship between smoking and the subsequent risk of developing lung cancer, a *cohort* of 1000 people were followed and the distribution was as follows:

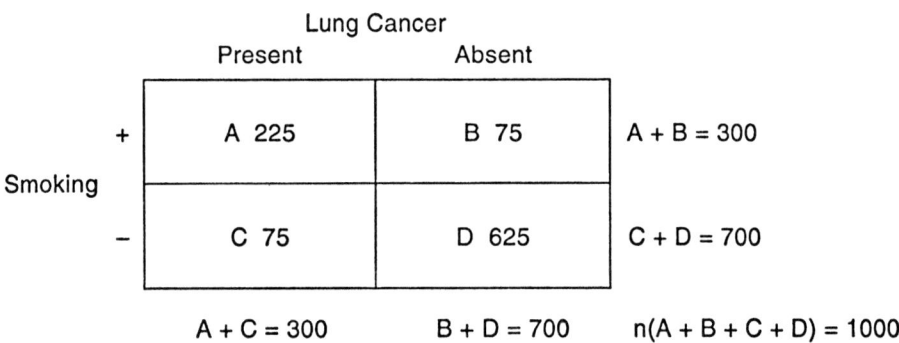

	Lung Cancer Present	Lung Cancer Absent	
Smoking +	A 225	B 75	A + B = 300
Smoking −	C 75	D 625	C + D = 700
	A + C = 300	B + D = 700	n(A + B + C + D) = 1000

$$\text{Incidence rate among smokers (absolute risk—see page 45)} = \frac{A}{A+B} = \frac{225}{300}$$

$$\text{Incidence rate among nonsmokers (absolute risk)} = \frac{C}{C+D} = \frac{75}{700}$$

As can be seen from the 2-by-2 table, 225 out of 300 smokers were eventually found to have lung cancer, as opposed to 75 of the 700 nonsmokers in this cohort study. The **relative risk** for smokers was then determined by the following ratio:

$$\text{The relative risk (in a } \textit{cohort study}\text{)} = \frac{\text{Incidence rate among } \textit{risk group}}{\text{Incidence rate among } \textit{non-risk group}}$$

$$\frac{A/(A+B)}{C/(C+D)} = \frac{225/(225+75)}{75/(75+625)} = \frac{0.750}{0.107} = 7$$

* **Further discussed on page 45.**

Therefore, relative to nonsmokers, smokers are *seven times as likely* to develop lung cancer based upon this *prospective (cohort) study*.

Warning: If, instead of being given a 2-by-2 table, you are given only the number of cases exposed to a risk factor [*A* of the 2-by-2 table] and the number of cases not exposed (*C* of the 2-by-2 table), the relative risk becomes a simple ratio of *A/C*, where

$$\frac{A}{C} = \frac{\text{Number of cases exposed to the risk factor}}{\text{Number of cases not exposed to the risk factor}}$$

Warning: If you are given only the number of cases exposed to a risk factor [*A* of the 2-by-2 table] and the total number of cases [*A* + *C* of the 2-by-2 table], simply subtract the number of cases exposed to the risk factor [*A*] from the total number of cases [*A* + *C*] to obtain the number of cases not exposed (*C* of the 2-by-2 table], and use the same formula of *A/C*.

Always remember, there are several ways to ask the same question—be careful!

A **relative risk** of greater than 1 is considered to be clinically significant.

Example

In a prospective study of the relationship between oral contraceptive use and the subsequent risk of developing endometrial cancer, a cohort of 1000 women were followed for 5 years. The results were as follows:

	Endometrial Cancer Present	Endometrial Cancer Absent	
OC +	A 245	B 75	A + B = 320
OC −	C 50	D 630	C + D = 680
	A + C = 295	B + D = 705	n = 1000

OC +   Using oral contraceptives
OC −   Not using oral contraceptives

Q   45. What was the incidence rate (absolute risk) of endometrial cancer among women who used oral contraceptives?

(A) 630/(50 + 630)
(B) 75/(245 + 75)

(C) 50/(630 + 50)

(D) 245/(245 + 75)

(E) Insufficient data

Q 46. What is the incidence rate (absolute risk) of endometrial cancer among women who didn't use oral contraceptives?

(A) 630/(50 + 630)

(B) 75/(245 + 75)

(C) 50/(50 + 630)

(D) 245/(245 + 75)

(E) Insufficient data

Q 47. What is the relative risk in this study?

(A) [75/(245 + 75)] / [50/(50 + 630)]

(B) [75/(245 + 75)] / [630/(50 + 630)]

(C) [50/(245 + 50)] / [630/(75 + 630)]

(D) [245/(245 + 75)] / [50/(50 + 630)]

(E) Insufficient data

Q 48. What is the incidence rate (absolute risk) of endometrial cancer among women who used oral contraceptives in person-years?

(A) $630/(680 \times 5)$

(B) $75/(320 \times 5))$

(C) $50/(630 \times 5)$

(D) $75/(320 \times 5)$

(E) $245/(320 \times 5)$

Q 49. Among 10 women with cervical cancer, medical records confirm a past history of herpes simplex type II infection in eight. What is the relative risk of developing cervical cancer in women with a history of HSV type II infection?

(A) 8/10

(B) 10/8

(C) 8/2

(D) 2/10

(E) 2/8

6.3   Retrospective or Case-Control Study*

In a **retrospective** or **case-control study,** researchers use documented medical records to select subjects *with a disease* **(cases)** and compare them to subjects *without the disease* **(controls)** to study differences between the two groups. These studies are usually *hospital-based* and are *easier, less time-consuming,* and *less expensive* than prospective studies, and require a *smaller number*

---

* This study attempts to answer the question, "What happened in the past?"

of subjects. In a case-control study, it is **not possible to determine** either an **incidence rate** or a **relative risk** because of the retrospective essence of this type of investigation as opposed to the prospective nature of incidence rate and risk determinations). This study is also *more prone to selection bias* than prospective (cohort) studies (particularly with respect to the selection of control groups), and is more appropriate for the study of *rare diseases*.

Example

A **case-control study** was done to investigate the relationship between oral contraceptive use and the subsequent risk of developing deep vein thrombosis among 110 women between 35 and 65 years of age. Cases were selected from hospital-based medical records with a confirmed diagnosis of deep vein thrombosis, and were then compared to controls who had no history of contraceptive use during their lifetime. The results of the study were as follows:

		Deep Vein Thrombosis		
		Present	Absent	
	+	A 40	B 20	A + C
OC				
	–	C 20	D 30	C + D
		A + C	B + D	

OC +   History of oral contraceptive use
OC –   No history of oral contraceptive use

6.4   Odds Ratio

Since we cannot determine an **incidence rate** or an accurate **relative risk** in a **retrospective case-control study,** how are we to estimate the *risk* of developing deep vein thrombosis either for the group who used oral contraceptives or the group who did not?

Because **case-control studies** are used for **rare diseases** (having a low incidence), if there is an *absence* of **selection bias** and **information (misclassification) bias** (see pages 43 and 44) with respect to the selected *control group* populations, an alternative is often used to retrospectively estimate the relative risk that existed for each group. This *estimate of relative risk* is referred to as the **odds ratio.**

$$\text{Odds ratio} = \frac{A \times D}{B \times C}$$

The **odds ratio** in this study is:

$$\frac{A \times D}{B \times C} = \frac{40 \times 30}{20 \times 20} = \frac{1200}{400} = 3$$

Therefore, the **odds ratio** states that, according to this *retrospective case-control study,* women with a history of oral contraceptive use were three times as likely to develop deep vein thrombosis as women without this history of birth control.

Q 50. Among a total of 52 women with deep vein thrombosis, medical records confirm that a total of 35 women used oral contraception during their lifetime. The odds ratio may be represented by

(A) 35/52
(B) 17/52
(C) 52/17
(D) 35/17
(E) 52/35

> **Warning:** As we have seen on page 37, there are several ways to ask the same question (see warnings concerning relative risk). Because cells B and D of the 2-by-2 table cannot be determined with the information given in the question, answer (C) represents the *best single answer* among the choices given. **Odds ratio** represents the *best estimate* of **relative risk**.

An **odds ratio** of greater than 1 is considered to have been significant in the absence of experimental error.

TABLE 6.1 COMPARISON BETWEEN COHORT AND CASE-CONTROL STUDY DESIGNS

CHARACTERISTICS	COHORT	CASE-CONTROL
synonyms	prospective, longitudinal, incidence	retrospective
question	what should happen	what has happened
view in 2-by-2 table	horizontal (rows)	vertical (columns)
onset of disease prior to study	no	yes
risk measurements	incidence, relative risk	odds ratio
usual site of study	community-based	hospital-based
types of diseases	relatively common	relatively rare
sample size	larger	smaller
cost	expensive	less expensive
time efficiency	very time-consuming	less time-consuming
difficulty level	tedious	relatively easy
selection bias	less likely	more likely

6.5 Cross-Sectional or Prevalence Study*

As the name suggests, these studies determine the **prevalence,** and *not the* **incidence,** of a disease. Because all subjects represent **existing cases,** both the *disease* and the *risk factor (attribute)* may be ascertained at the *same time.* This type of study will describe both components at one *point in time* (**point prevalence**) or during a specified *period of time* (**period prevalence**).

These studies are *easy, quickly performed,* and relatively *inexpensive.* **Cross-sectional studies** do not necessarily establish *causal relationships.* Surveys and polls are *cross-sectional* in nature.

Example

To minimize morbidity and mortality resulting from stroke (cerebrovascular accident) secondary to hypertension, the *prevalence* of stroke and the blood pressures of a sample of county residents over the age of 35 were recorded in 1991 to justify the inception of a stroke prevention program in Orange County, California.

Example

Upon determining the *prevalence* of coronary heart disease and its association with high levels of cholesterol, blood samples were taken from high school teenagers in Dade County, Florida, in April 1992 to record serum cholesterol levels in order to implement a new nutritional program of a low-fat, high-fiber diet in the public school lunch system.

6.6 Experimental Studies or Randomized Clinical Trials

The purpose of these studies in clinical medicine is to determine which treatment is superior among competing treatments. This involves **randomization** of patients into various groups, and minimizes the potential for **selection (sampling) bias.** The study is *prospective* in nature—that is, subjects are followed over a period of time. This study is most commonly referred to as a **randomized clinical trial.**

In **double-blind clinical trials,** both the subjects and the investigator(s) are blinded—that is, neither knows into which groups the subjects are enlisted. This study design reduces the potential for *selection (sampling) bias.*

In **cross-over studies,** one group is given a specific treatment, and the other group a placebo. After a specified period of time, the assignment is reversed. This study design also minimizes the potential for *selection bias.*

* This study attempts to answer the question, "What is happening right now?"

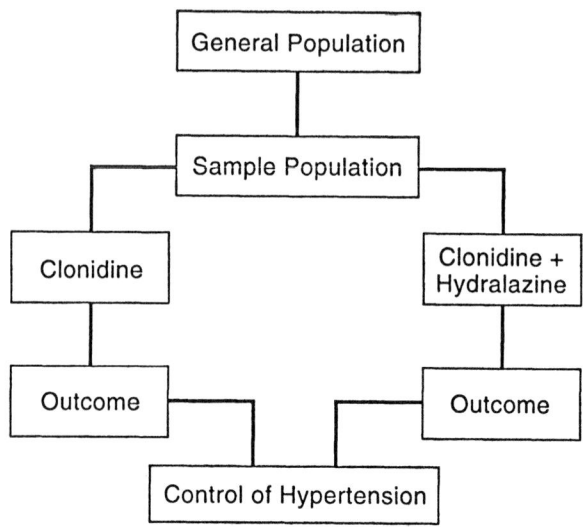

Fig. 6.1 Experimental Study.

6.7   Independent Variables

An **independent variable** is one that the researcher can either introduce or isolate in order to demonstrate its *effect upon* a **dependent variable.** In the following study, the independent variable is *smoking.*

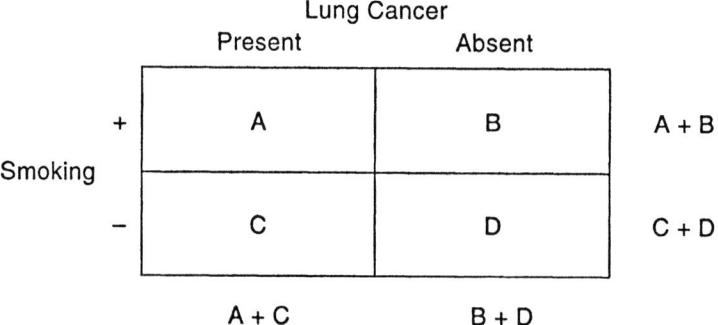

6.8   Dependent Variables

A **dependent variable** is one that may be *present, absent,* or *altered* when an **independent variable** is *present, absent,* or *altered.* In the preceding study, the dependent variable was *lung cancer.*

6.9   Confounding Variables*

A **confounding variable** is one that *affects both* the *dependent* and *independent variables*—that is, has an association with both

* Further discussed on page 45.

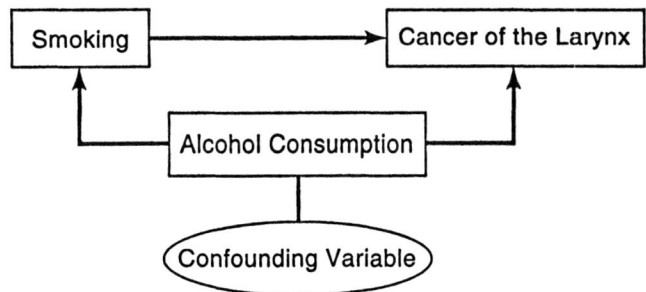

Fig. 6.2 Confounding Variable.

the disease and the risk factor under study that may distort relationships between the two and *confound* the study results.

Example

In a *case-control study* assessing the relationship between smoking and the subsequent risk of developing laryngeal cancer, **alcohol consumption** may be a *confounding variable*.

Cigarette smokers are often drinkers of alcohol. Because alcohol consumption has been shown to increase both the frequency of cigarette smoking and the risk of developing laryngeal cancer, it is extremely important to establish *controls* for alcohol as a potential *confounding variable* in the study analysis. Establishing **controls** for the confounding variable (alcohol) allows us to demonstrate a true statistical association between the dependent and independent variables of laryngeal cancer and smoking, respectively.

Confounding variables may not always be as easy to identify as in the preceding example. The possibility of the existence of unrecognized confounding variables must always be considered in experimental studies.

6.10 Bias (Systemic Error)

As was noted earlier in Chapter 3, the *validity* of a test is dependent upon the *accuracy* of test classifications and measurements. When there is a distortion of a test measurement that results in a unidirectional deviation from the mean that cannot be corrected by statistical manipulation, it is referred to as a **nonrandom systemic error,** or **bias.**

Three of the most problematic forms of bias in medicine are: **selection (sampling) bias, information (misclassification) bias, and confounding.**

Selection (Sampling) Bias: Selection bias occurs when study results become distorted by the selection process. This distortion may occur in many different ways. Some of the important ones are the following.

Admission rate (Berkson's) bias: Distortions in risk ratios occur as a result of *different hospital admission rates* among cases with the risk factor, cases without the risk factor, and controls with the risk factor—causing greatly different risk-factor probabilities to interfere with the outcome of interest. This type of bias can be reduced by choosing controls from a wide variety of disease categories—that is, *randomization.*

Nonresponse bias: A common problem encountered in household health surveys is the noncompliance of people who have scheduled interviews at their homes. The most valid way to manage the problem of "nonresponse" is to try repeatedly to visit or call the nonresponders at their homes. If this is unsuccessful, the most appropriate way to manage the nonresponsive subjects is to include them in the survey but treat them as *unknown* in the data analysis.

Lead time bias: Very often a *time differential* exists between diagnosis and treatment among sample subjects, which may result in higher survival rates being erroneously attributed to superior treatment rather than to early detection. An appreciable time differential may also exist between diagnosis and the onset of disease within a given sample, producing artificially low incidences for a given period.

□ NOTE

Selection (sampling) bias is only the **nonrandom (systemic)** component of sampling error. Errors in sampling are also caused by **random error**—that is, random variation within a sample that is strictly attributable to *chance* rather than to a sample that is unrepresentative of the general population (*bias*). **Bias** in a study may be reduced by *increasing sample size* and by *equalizing the chances of each member of a population to be chosen* for the sample. This is what is known as **randomization.**

Information (Misclassification) Bias: Information bias occurs when study results become distorted by poor data collection or inaccurate measurements of variables. Some of the more common forms of this type of bias are the following.

Recall bias: Differentials in the memory capabilities of sample subjects may cause risk-factor exposures to be under- or overreported.

Interviewer bias: Because the "blinding" of interviewers to diseased and control subjects is often difficult, subject responses may be influenced by variations in the interviewer's tone of voice, body language, probe level, and perceived preference level—all of which may be influenced by the interviewer's perception of the subjects' condition

Unacceptability bias: Patients often reply to an interviewer's questions with "desirable" answers regarding dietary, drug, exercise, behavioral, and recreational habits, resulting in understated measurements of many risk factors and other pertinent variables.

Confounding: As was discussed on page 42, a *confounding variable* may lead to bias in a study because it has a relationship with both the dependent and independent variables that either masks or potentiates the effect of the variable under study.

B. Measures of Risk

Factors that are likely to increase the *incidence, prevalence, morbidity*, or *mortality* of a disease are called **risk factors.** Three valuable estimates are used to measure risk: **absolute risk, relative risk,** and **attributable risk.**

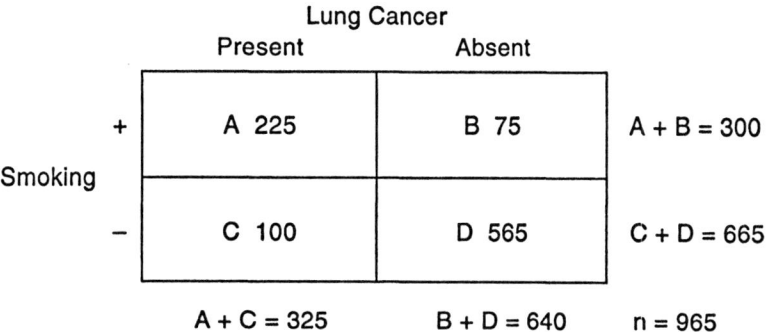

6.11 Absolute Risk

Absolute risk allows us to separately calculate the incidences of a particular disease in both populations of a risk factor study for the purpose of making individual risk comparisons for each population. Absolute risk may be determined for the population of people exposed to a risk factor as well as for those not exposed to the risk factor.

The Absolute Risk for smokers $= A/(A+B)$
$$= 225/300 = 0.75 = 75\%$$

The Absolute Risk for nonsmokers $= C/(C+D)$
$$= 100/665 = 0.15 = 15\%$$

Therefore, 75% of this population of smokers eventually developed lung cancer as opposed to only 15% of the study's nonsmoking population.

6.12 Relative Risk

Relative risk gives us risk as a *ratio* of the incidence among subjects exposed to a particular risk factor divided by the incidence among subjects who were not exposed to the risk factor. Calculations of high-risk-group incidences *relative to* disease incidences in the general (average risk) population are one of the most important ratios used in clinical and preventative medicine.

$$\frac{\text{Relative}}{\text{Risk}} = \frac{\text{Incidence rate among those exposed to the risk factor}}{\text{Incidence rate among those not exposed to the risk factor}}$$

$$= \frac{A/(A+B)}{C/(C+D)} = \frac{225/(225+75)}{100/(100+565)} = \frac{0.750}{0.150} = 5$$

Therefore, relative to nonsmokers, smokers are five times as likely to develop lung cancer, based upon this study.

☐ NOTE

A relative risk of greater than one is always important in the clinical evaluation of a patient.

6.13   Attributable Risk

Attributable risk allows us to attribute *differences* in the incidences of a disease to a particular risk factor. This is done by simply subtracting the incidence among those not exposed to a risk factor from the incidence among those who were exposed.

$$\frac{\text{Attributable}}{\text{risk}} = \frac{\text{Incidence rate among those}}{\text{exposed to the risk factor}} - \frac{\text{Incidence rate among those}}{\text{not exposed to the risk factor}}$$

Attributable risk $(AR) = A/(A+B) - C/(C+D)$

$AR = A/(A+B)$ {Incidence among exposed} $- C/(C+D)$ {Incidence among nonexposed}

$$= 225/(225+75) - 100/(100+565) = 0.750 - 0.15 = 0.60$$

Often, *attributable risk* is expressed as **attributable risk percent,** where **attributable risk** is a *percentage of the* **absolute risk** (incidence rate among those exposed to the risk factor—for example, smoking).

$$\text{Attributable risk percent} = \frac{\text{Attributable risk}}{\text{Absolute risk}} \times 100$$

$$= \frac{\text{Attributable risk}}{\text{Absolute risk (smokers)}} \times 100$$

$$= \frac{0.60}{0.75} \times 100 = 80\%$$

Therefore, 80% of the time, the differences (variations) in the incidence of lung cancer between those exposed to the risk factor (smoking) and those not exposed to the risk factor may be directly attributable to the presence of the risk factor in this particular study. This percentage is most frequently used to justify the inception of risk prevention programs when attributable risk factor percentages are deemed to be high.

Warning: The preceding estimates (relative risk, attributable risk, and attributable risk percent), as with all others that we have discussed, do not necessarily establish a *cause and effect* relationship between risk factors and disease. They do, however, support the hypotheses made in many reputable studies throughout the medical community that risk factors may be considered to be contributory factors to specific diseases, and may have either a direct or indirect influence on their incidence, prevalence, morbidity, and/or mortality.

Summary

	Lung Cancer		
	Present	Absent	
Smoking +	A 225	B 75	A + B = 300
Smoking −	C 100	D 565	C + D = 665
	A + C = 325	B + D = 640	n = 965

Absolute risk $= A/(A+B) = 225/(225+75) = 0.75 = 75\%$

Relative risk $= \dfrac{A/(A+B)}{C/(C+D)} = \dfrac{225/(225+75)}{100/(100+565)} = \dfrac{0.750}{0.150} = 5$

Attributable risk $= A/(A+B) - C/(C+D) = 0.75 - 0.60$

Attributable risk percent $= \dfrac{\text{Attributable risk}}{\text{Absolute risk}} \times 100 = \dfrac{0.60}{0.75} \times 100 = 80\%$

Example

In a *prospective study* of the relationship between HIV status and the subsequent risk of developing non-Hodgkin's B-cell lymphoma, a cohort of 600 men from New York City were followed from 1981 to 1991. The results were as follows:

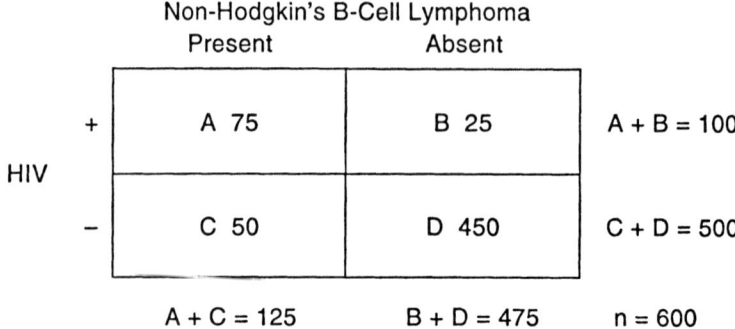

$$\text{Absolute risk} \quad = A/(A + B) = 75/(75 + 25) = 0.75 = 75\%$$

$$\text{Relative risk} \quad = \frac{A/(A + B)}{C/(C + D)} = \frac{75/(75 + 25)}{50/(50 + 450)} = \frac{0.750}{0.100} = 7.5$$

$$\text{Attributable risk} = A/(A + B) - C/(C + D)$$
$$= 75/(75 + 25) - 50(50 + 450) = 0.75 - 0.10$$
$$= 0.65$$

$$\frac{\text{Attributable risk}}{\text{percent}} = \frac{\text{Attributable risk}}{\text{Absolute risk}} \times 100 = \frac{0.65}{0.75} \times 100 = 87\%$$

The Normal (Gaussian) Distribution

Descriptive statistical measurements are often used in medical literature to summarize data. Two parameters that are most frequently used in clinical medicine are *measures of central tendency* and *measures of dispersion.*

7.1 Measures of Central Tendency

A measure that describes a *typical value* in a set of data is referred to as a **measure of central tendency.** Three measures of central tendency describe such values when they are found in a normally distributed sample: **mean, median,** and **mode.**

   Mean: the sum of the scores divided by the number of scores—that is, the average score.

   Median: the 50th percentile or midpoint of a sequence—that is, the score above which, and below which, half of the scores are found.

   Mode: the score that occurs most frequently.

☐ NOTE

Although all measures of central tendency may be altered by the addition of very high or very low values in a distribution, the mode is usually unaffected by such values, and the mean is susceptible to the greatest degree of change.

Example

In a follow-up study of five patients admitted to the coronary care unit with a diagnosis of acute myocardial infarction, the length of stay was found to be 5, 3, 8, 5 and 9 days. Calculate the mean, median, and mode for this sample of patients.

   First arrange the data in ascending or descending order:

   3, 5, 5, 8, 9

   Mean = Sum of the scores / Number of scores
        $= 3 + 5 + 5 + 8 + 9 = 30 / 5 = 6$

Median = Midpoint of the sequence = 5

(If there is an even number of values, the mean of the middle two numbers becomes the median.)

Mode = The score that occurs most frequently = 5

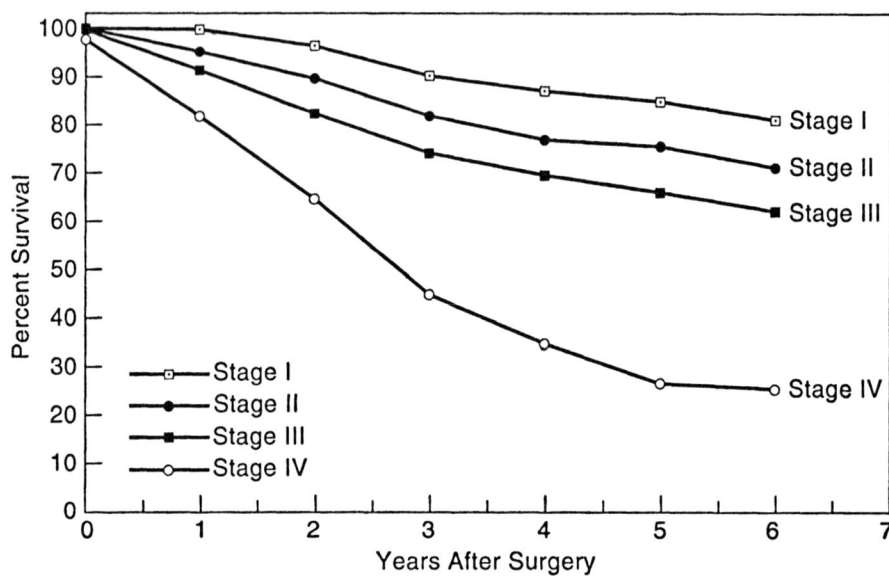

Fig. 7.1 Survival Rates of Patients with Astrocytoma.

Refer to Figure 7.1 to answer the following questions.

Q   51. The four-year survival rate for a patient with Stage III astrocytoma in this study is

(A) 20%
(B) 40%
(C) 60%
(D) 70%
(E) 100%

Explanation: Locate the four-year survival point on the x axis, draw a straight line to intersect the Stage III curve, and then connect this point to the y axis, which indicates a 70% survival rate (choice D).

Q   52. The median survival for patients with Stage IV astrocytoma in this study is

(A) 1.5 years
(B) 2.5 years
(C) 3.5 years
(D) 4.5 years
(E) 5.5 years

Explanation: This procedure is just the opposite of what was done previously. Instead of looking first at the x axis, look for the 50% (median) survival point on the y axis. From this point, draw a straight line to intersect the Stage IV curve, and then connect this point with another straight line to the x axis, which represents 2.5 years after surgery (choice B).

7.2 Measures of Dispersion

A measure that describes the spread or variation of the observations is referred to as a **measure of dispersion.** Four measures of dispersion that are used in clinical epidemiology and medicine are the **range, standard deviation, variance,** and **standard error of the mean.**

Range: the difference between the highest and lowest observations.

Standard deviation: A measure of the *spread or dispersion of the values around the mean.* The standard deviation gives us an indication of how much variability can be expected among the scores. Although you will not be asked to calculate a standard deviation on the board examinations, it is the square root of the sum of the squares of each score's dispersion from the mean, divided by the sample size, minus one. It is calculated by using the formula:

$$s = \sqrt{\frac{\sum(x - \bar{x})^2}{n - 1}}$$

Where: x = sample score, $\bar{x}$ = mean of the sample, and
n = sample size

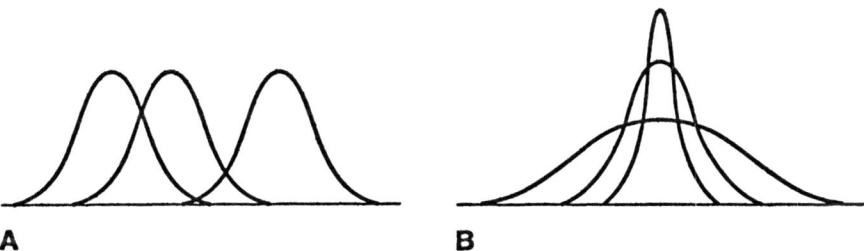

A

B

(a) same standard deviations, with different means

(b) same means, with different standard deviations

Fig. 7.2 Central Tendency and Dispersion.

Variance: An additional measure of the spread or dispersion of the values around the mean. Simply stated, variance is the *square of the* **standard deviation.**

$$s^2 = \frac{\sum(x - \bar{x})^2}{n - 1}$$

Standard error of the mean (SEM or SE{x}): The standard error of the mean plays an important role in many of the statistical procedures used in epidemiology and clinical medicine. It is used for confidence limit determinations and becomes an *estimate of the standard deviation of the population* through the following formula:

$$SEM = \frac{s}{\sqrt{n}}*$$

Where: s = standard deviation of the sample
n = sample size

☐ NOTE

You *will not* be asked to calculate a *standard deviation, variance,* or *standard error of the mean* on the examination.

7.3   The Normal Distribution

The normal (Gaussian) distribution is a theoretical binomial probability distribution that is represented by a bell-shaped polygon called a normal curve which is symmetrical about a point represented by the *mean, median,* and *mode.*

Characteristics of the Normal Distribution Curve

It is bell-shaped.
It has symmetry around the mean.
The mean, median, and mode are equal.
The dispersion or spread from the mean is represented by the standard deviation.
68% (two-thirds) of the values fall within one standard deviation of the mean.
95% of the values are found within two standard deviations of the mean.
99% of the values are found within three standard deviations of the mean.

---

* As you can see, the standard error of the mean decreases as the size of the sample increases.

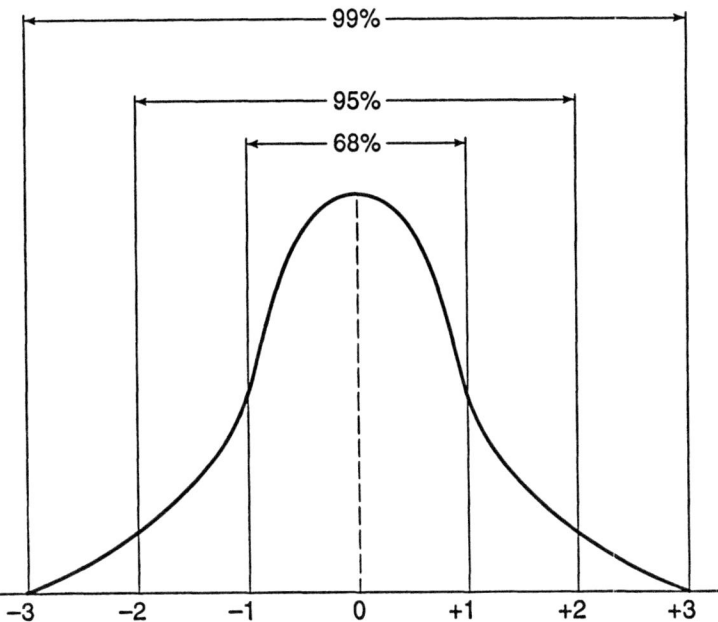

Number of Standard Deviations (s) from the Mean

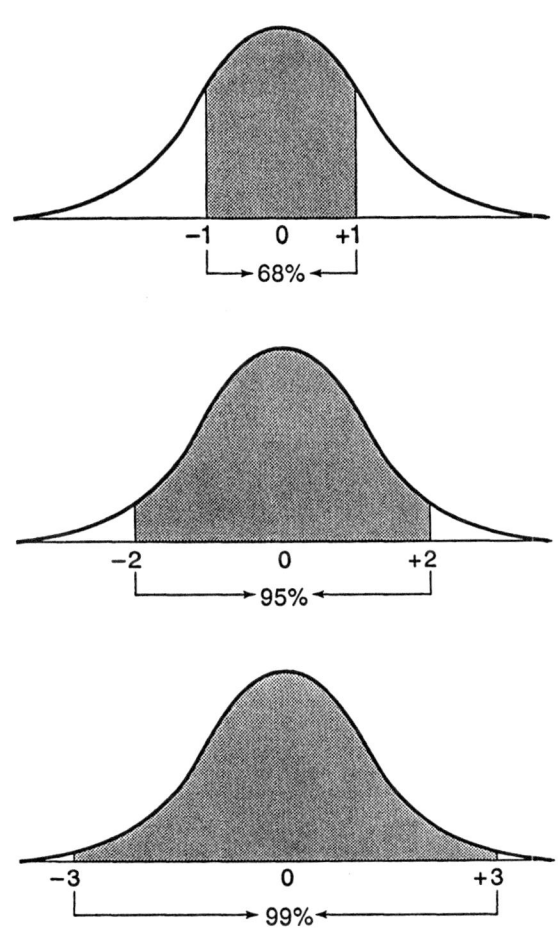

Fig. 7.3 The Normal (Gaussian) Distribution.

Q 53. The characteristics of the normal distribution curve include the following:

(A) The total area under the curve represents 100% of all values.
(B) The mean and median are found below the apex.
(C) 5% of the values lie beyond 2 standard deviations from the median.
(D) The curve is perfectly symmetrical.
(E) All the above.

Q 54. In a study at the University of Alabama at Tuscaloosa, the mean systolic blood pressure of 250 medical students was 116 mm Hg, with a standard deviation of 4 mm Hg. From the data, 99% of the medical students will have systolic blood pressures (mm Hg) in the range of

(A) 110–130
(B) 104–128
(C) 112–120
(D) 116–124
(E) 118–122

Q 55. In a study involving 150 nursing students at the University of Nevada at Las Vegas, the mean serum cholesterol level was found to be 176 mg/dL, with a sample variance of 25 mg/dL. From the data, one-third of the nursing students will not have a cholesterol level (mg/dL) in the range of

(A) 161–191
(B) 166–186
(C) 171–181
(D) 172–180
(E) 175–177

Q 56. From the data, 95% of the nursing students will have a cholesterol level in the range (mg/dL) of

(A) 161–191
(B) 166–186
(C) 171–181
(D) 172–180
(E) 175–177

7.4   Skewed (Asymmetric) Distributions

When the left- and right-hand sides of a frequency distribution do not approximate mirror images, the data are said to be skewed or asymmetrical.

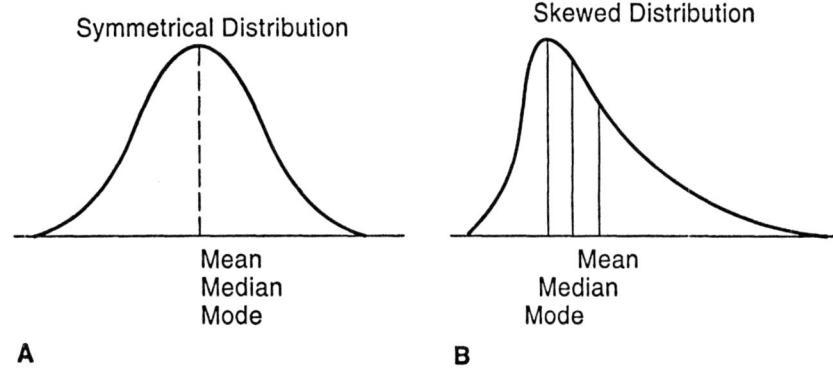

Fig. 7.4 Symmetrical and Skewed Distributions.

The direction of the tail of the curve indicates the direction of the skewed distribution. If the *tail* of the curve is toward the *right,* the distribution is said to be *positively skewed;* if the tail of the curve is toward the *left,* the distribution is *negatively skewed.*

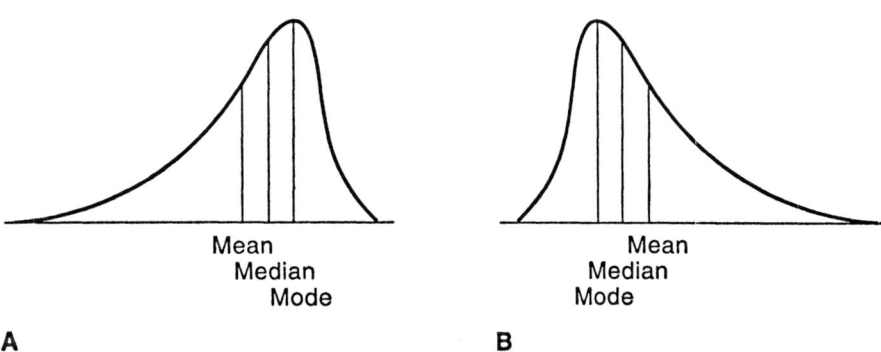

Fig. 7.5 Positive and Negative Skewness.

NEGATIVELY SKEWED	POSITIVELY SKEWED
skewed toward the **left**	skewed toward the **right**
mean **less than** median	mean **greater than** median

☐ NOTE

In a **skewed distribution,** the *mean* always follows the *tail* of the curve. From the tail of the curve to the apex (mode), the *mean, median,* and *mode* are always in **alphabetical order.** Remembering this little "pearl" will negate the possibility of answering questions about skewed distributions incorrectly.

Example

In a study of 150 Native Americans in Arizona, serum cholesterol levels were followed to determine their association with coronary artery disease. The results of the study were as follows:

CHOLESTEROL LEVELS
Mean   = 219 mg/dL
Median = 199 mg/dL
Mode   = 159 mg/dL

Q 57. From the data, it can be concluded that

    (A) the distribution is normal
    (B) the distribution is negatively skewed
    (C) the distribution is positively skewed
    (D) the distribution is skewed toward the left
    (E) there are insufficient data to determine the distribution

7.5   Summary of Distributions

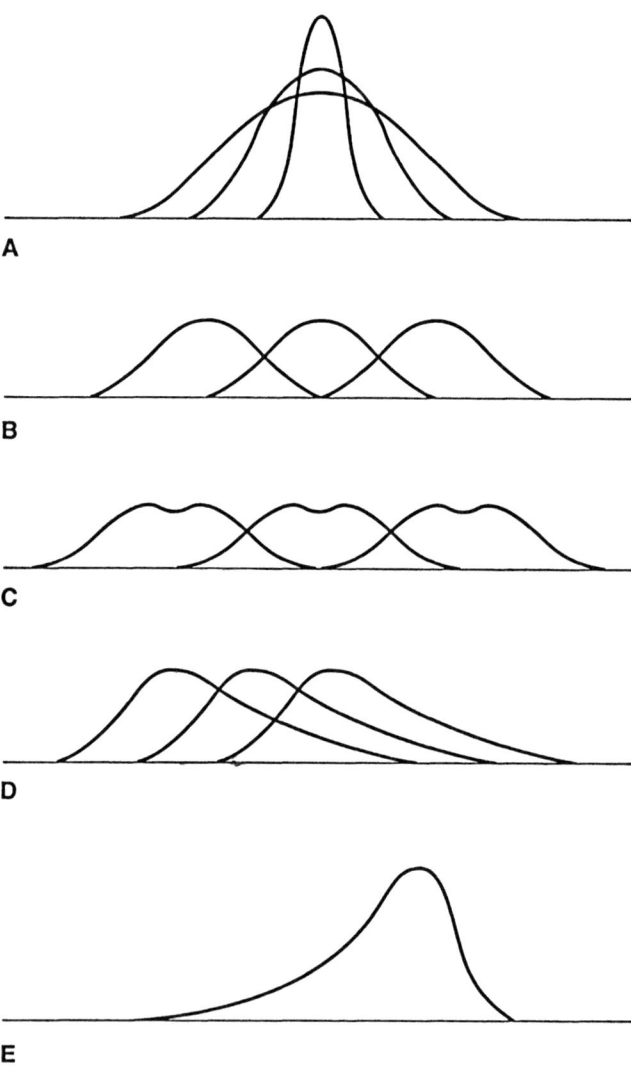

Fig. 7.6 Distribution Summary.

A. Same means with different standard deviations
B. Same standard deviations with different means
C. Bimodal distributions
D. Positively skewed distributions
E. Negatively skewed distribution

CHAPTER 8

Statistical Tests

8.1 Student's *t* Test

When is a student's *t* test used?

The student's **t test** is used to *compare the means of two small* (*n < 30*) *independent samples* for the purpose of determining the statistical significance (*p* value) of the observed findings (as opposed to the *F test,* which compares three or more means). The **p value** represents the *probability that the results occurred* **purely by chance,** rather than as a result of the variable under study.

Three components are needed to determine the *statistical significance* of an observed finding:

A. *t value* (critical value): $t = \dfrac{\bar{x} - \bar{\mu}}{s / \sqrt{n}}$

Where: $\bar{x}$ = sample mean
$\bar{\mu}$ = mean of the standard population
s = standard deviation
n = sample size (< 30 for *t* tests)

B. Degree of freedom (*df*) = sample size $-1 = n - 1$
C. Standard *t* distribution table (lists *t* values and degrees of freedom with their corresponding *p* values)

Example

Temperatures of 26 patients were recorded 48 hours after surgery. The mean temperature of this group was found to be 99.1°F, with a standard deviation of 1.0°F. The standard (normal) temperature is 98.6°F. The chief resident was asked to show other surgical residents whether or not there was a statistically significant difference between the temperatures of surgical patients 48 hours postoperatively and the standard (normal) temperature of 98.6°F.

HOSPITAL DATA

Sample mean temperature ($\bar{x}$)	= 99.1°F
Standard temperature ($\bar{\mu}$)	= 98.6°F
Standard deviation (*s*)	= 1.0°F
Sample size (*n*)	= 26

* **Note: For larger sample sizes (n > 30), z scores are used.**

To determine whether the temperatures of the postoperative patients were significantly different from the normal temperature of 98.6°F, the resident must determine the following values:

A. *t value*
B. degree of freedom (d_f)
C. *p* value

Calculations are as follows:

A. $t \text{ value} = \dfrac{\bar{x} - \bar{\mu}}{s / \sqrt{n}} = \dfrac{99.1 - 98.6}{1.0 / \sqrt{26}} = \dfrac{0.050}{0.196} = 2.55$

B. degree of freedom $(d_f) = n - 1 = 26 - 1 = 25$
C. *p* value: Step 1. Locate the row (horizontal) that corresponds to 25 degrees of freedom. Step 2. Locate the column (vertical) that corresponds to $t \leq 2.55$. Step 3. Locate the *p* value at the top of that column.

STUDENTS' *t* DISTRIBUTION TABLE

DEGREE OF FREEDOM	PROBABILITY (*p* value)			
	0.50	0.10	0.05	0.01
1	1.000	6.31	12.71	63.66
5	0.727	2.02	2.57	4.03
10	0.700	1.81	2.23	3.17
20	0.687	1.71	2.06	2.79
25	0.684	1.71	2.06	2.79
Infinity	0.674	1.64	1.96	2.58

As you can see, $p < 0.05$

WARNING 1: After locating the row corresponding to the appropriate degree of freedom (d_f), do not pick the *t* value in that row that is *closest* to the calculated *t* value! Pick the *t* value that is *less than* (or equal to) the given *t* value.

WARNING 2: The comprehensive medical examinations *may not label the columns and rows* of the standard *t distribution table*. You must know that the first column on the left is for *degrees of freedom* (d_f); the top row is for *p values* (level of significance); and the remainder of the table lists the *t values*.

Conclusion: There is a statistically significant difference between the temperatures of postoperative patients and the standard mean (normal) temperature of 98.6°F. A *p value* that is < 0.05 is always considered *significant*.

☐ NOTE

You *will not* be asked to calculate a *t* value, but *you must know how to determine the significance level (p value)* on the examination by using the *t* table given for the question.

Example

A dietary study was conducted to measure levels of serum cholesterol to determine the possible risk of developing coronary heart disease. The *mean* and *standard deviation* were computed for two groups of medical students, and the results were as follows:

	MEAN	STANDARD DEVIATION
Group 1 = 176		25 (mg/dL)
Group 2 = 225		35 (mg/dL)
t value = 2.78		

Degrees of freedom $(d_f) = 25$

STUDENTS' *t* DISTRIBUTION TABLE

DEGREE OF FREEDOM	PROBABILITY (*p* value)			
	0.50	0.10	0.05	0.01
1	1.000	6.31	12.71	63.66
5	0.727	2.02	2.57	4.03
10	0.700	1.81	2.23	3.17
20	0.687	1.71	2.06	2.79
25	0.684	1.71	2.06	2.79
Infinity	0.674	1.64	1.96	2.58

Q. 58. The corresponding *p* value is

(A) <0.50
(B) <0.10
(C) <0.05
(D) <0.01
(E) <0.001

8.2   *F* Test (ANOVA—Analysis of Variance)

When is an *F* test used?

The **F test** is used *to compare means of three or more samples or groups* for the purpose of determining the statistical significance of the observed findings (as opposed to the *t test*, which compares means of two samples or groups). The technique used to compare means of three or more groups is called **analysis of variance (ANOVA)**. Although *computations* concerning *F* tests are *not needed* for examination purposes, the *appropriateness* of its use in clinical medicine *must be understood*.

Q 59. The four blood groups A, B, O, and AB were studied to compare the quantitative serologic differences among their antigenic structures. The most appropriate statistical test to make this determination is a(n)

(A) *t* test
(B) *F* test (ANOVA)
(C) chi-square test
(D) correlation analysis
(E) regression analysis

8.3 The Chi-Square Test*

When is a *chi-square test* used?

The **chi-square (χ^2) test** is used for *comparing two or more independent proportions* within two or more groups, making it appropriate for multigroup comparisons (as opposed to comparing two means in a *t test*, or comparing three or more means in an *F test*). For examination purposes, if the data are arranged in a 2-by-2 table, you may use the *chi-square test*.

	Disease Present	Disease Absent	
Test +	A	B	A + B
Test −	C	D	C + D
	A + C	B + D	n = (A+B+C+D)

To determine the statistical significance of an observed finding, three steps are needed:

1. Chi-square value

$$\chi^2_{df} = \sum \frac{(O - E)^2}{E} \quad \text{or} \quad \frac{N(AD - BC)}{(A + B)(C + D)(A + C)(B + D)}$$
(without Yate's correction)

Where: O = observed number
E = expected number
$\sum$ = the sum of

The *expected number* for any cell (*A, B, C,* or *D*) can be generated by multiplying the corresponding total in that column [(*A* + *C*) or

* χ^2 is a Greek symbol that is pronounced *"ki-square."*

$(B + D)$] with the corresponding total in that row [$(A + B)$ or $(C + D)$ respectively], and then dividing the product by the total sample size $(A + B + C + D)$.

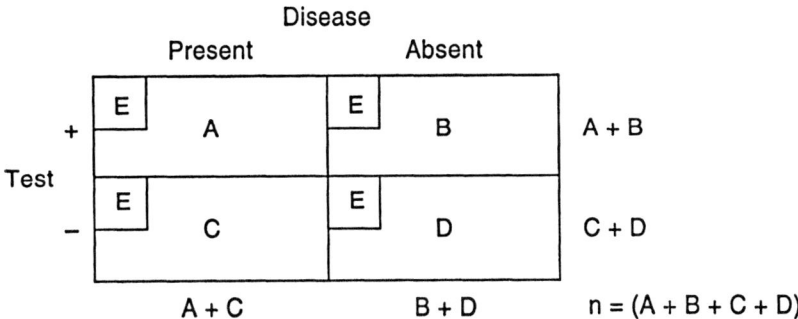

2. *Expected number* in cell A

$$E_A = \frac{(A+C) \times (A+B)}{(A+B+C+D)}$$

3. *Degree of freedom* (d_f)*

$$d_f = (\text{number of columns} - 1) \times (\text{number of rows} - 1)$$
$$= (c - 1) \times (r - 1)$$

4. The *standard χ^2 distribution table* lists χ^2 values and *degrees of freedom* with their corresponding *p values.*

E x a m p l e

To assess the possible association between 100% oxygen therapy and the subsequent development of retrolental fibroplasia, a total of 135 premature infants in the intensive care unit were studied. The results were as follows:

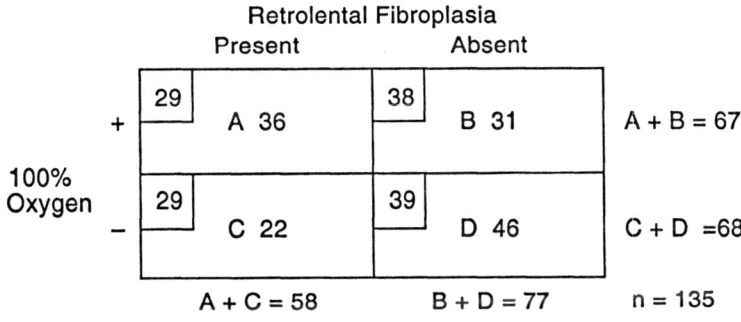

*For a 2-by-2 table, the d_f will be $(2 - 1) \times (2 - 1) = 1$.

Calculations are as follows:

CELL	OBSERVED	EXPECTED	$(O - E)$	$(O - E)^2$	$\dfrac{\chi^2}{(O - E)^2/E}$
A	36	28.785	7.215	52.215	1.808
B	31	38.215	−7.215	52.215	1.362
C	22	29.215	7.215	52.215	1.781
D	46	38.785	−7.215	52.215	1.342
				Total	6.293

1. χ^2 value = 6.293
2. Degree of freedom $d_f = (2 - 1) \times (2 - 1) = 1$

CHI-SQUARE DISTRIBUTION TABLE

DEGREE OF FREEDOM	PROBABILITY (p value)				
	0.99	0.95	0.05	0.01	0.001
1	0.000157	0.00393	3.841	6.63	10.83
2	0.0201	0.103	5.991	9.21	13.82
3	0.115	0.352	7.815	11.34	16.27
4	0.297	0.711	9.488	13.28	18.47
5	0.554	1.145	11.070	15.09	20.51

3. From the χ^2 distribution table, **$p < 0.05$.**

Conclusion: There is a significant difference in the incidence of retrolental fibroplasia in premature infants who received 100% oxygen when compared to those who did not.

☐ NOTE

You *will not* be asked to calculate a χ^2 value, but you will be asked to interpret the given χ^2 value as it relates to its *statistical significance*.

Warning: After locating the row corresponding to the appropriate degree of freedom (d_f), *do not* pick the χ^2 value in that row that is *closest* to the calculated χ_2 value! Pick the χ_2 value that is *less than* (*or equal to*) the given χ_2 value.

Warning: The comprehensive medical examinations *may not label* the columns and rows of the *standard chi-square distribution table.* You must know that the first column on the left is for degrees of freedom (d_f), the top row is for p values (level of significance), and the remainder of the table is for χ^2 values (chi-square critical values).

Correlation Coefficient (r) and Linear Regression Analysis

So far, we have seen how one could state whether or not a difference between two groups is statistically significant (Chapter 4). Such a comparison between two or more groups can be viewed as an examination of the association or relationship between two variables. The relationship between two variables (x and y) can be demonstrated by two methods:

Correlation Coefficient: Measures the strength of the association between two variables.

Regression Analysis: Provides an equation that estimates the change in a dependent variable (y) per unit change in an independent variable (x).

A.   Correlation   Coefficient

9.1   The correlation coefficient (r) measures the strength of the association between two variables (Pearson correlation).

The value of *r* ranges between +1 and –1.

+1 = A positive or direct correlation—eg, smoking and lung cancer

–1 = A negative or inverse correlation—eg, exercise and the risk of heart disease

0 = A zero correlation (no relationship)—eg, color of the skin and intelligence

☐ NOTE

Even with a perfect correlation (r) of +1 or –1, a causal relationship may not necessarily exist between the two variables.

The relationship between *two correlated variables* may be visualized using a graph called a **scattergram** (scatter diagram), as the following figure illustrates.

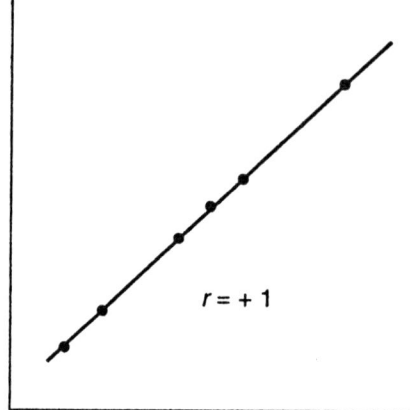

Perfect positive correlation

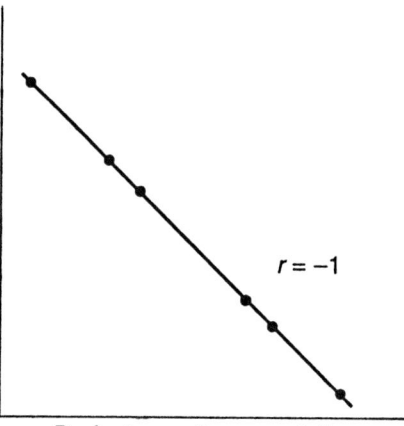

Perfect negative correlation

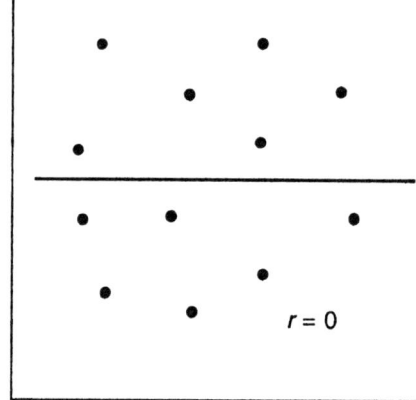

No correlation

Fig. 9.1 The Scatter Diagram.

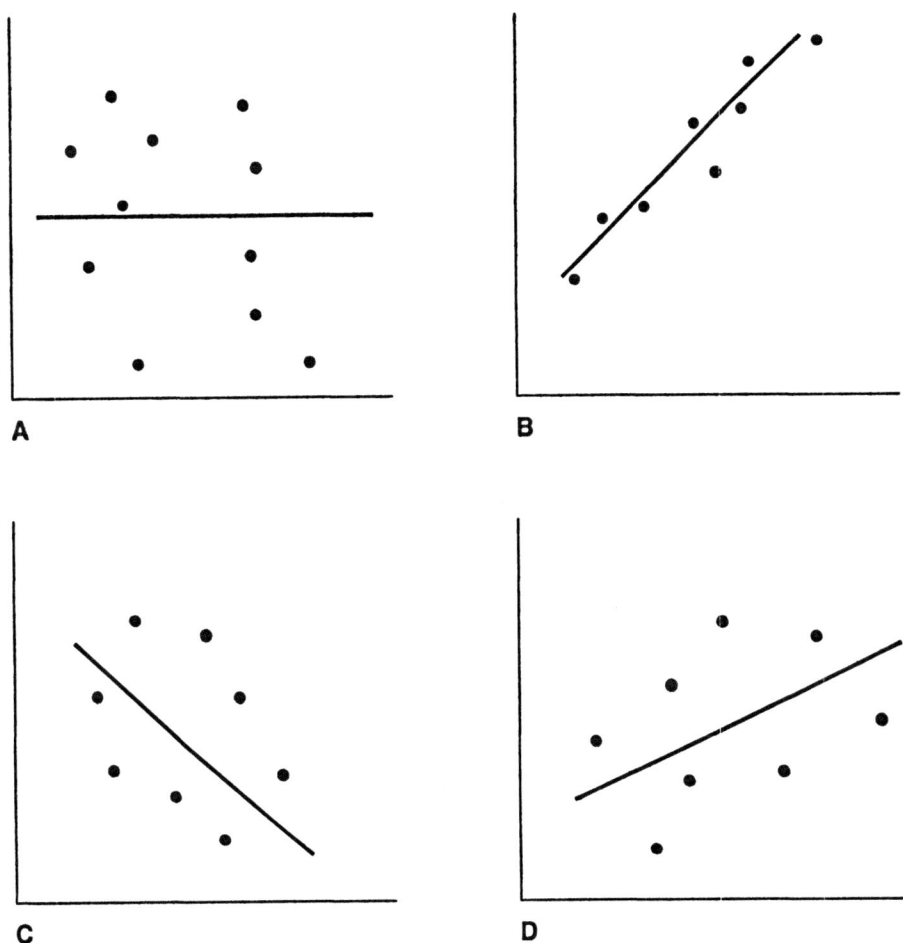

A. Correlation coefficient of zero
B. Positive correlation coefficient of high magnitude
C. Negative correlation coefficient of low magnitude
D. Positive correlation coefficient of low magnitude

Fig. 9.2 Correlation Coefficients.

9.2   Limitations of the Correlation Coefficient

Although **correlation analysis** is very helpful in determining the degree to which two variables are associated, it becomes very ineffective when the total number of observations in the study are small ($n < 30$). When small sample scattergrams are plotted, the *outliers* (observations that lie outside the normal range) have a marked effect on the correlation coefficient and often produce misleading results.

☐ NOTE

Regardless of the size of the sample or the magnitude of its correlation coefficient, the relationship between two correlated variables is *strictly* one of *association*. Remember, association is not synonymous with causality.

Q 60. Studying the association between plasma levels of renin and changes in blood pressure, a researcher would obtain the most effective use of the data by the application of:

(A) student's t test
(B) F test
(C) chi-square test
(D) analysis of variance
(E) correlation analysis

9.3 Coefficient of Determination

The **coefficient of determination (r^2)** is the *square of the* **correlation coefficient (r),** and represents the proportion of the total variation in a dependent variable that is determined (explained) by, or associated with, the independent variable.

Example

In a study of the association between diethylstilbestrol (DES) and the development of vaginal cancer, the researcher reported a *correlation coefficient (r)* of 0.91 or $r^2 = 82\%$. That is, 82% of the variation in the incidence of vaginal cancer (dependent variable) can be associated with diethylstilbestrol (independent variable). The remaining 18% of the variation cannot be explained, and may be due to other factors not considered in the study analysis.

Q 61. An investigator used a test for identifying people with ankylosing spondylitis that has a maximum intercorrelation (correlation coefficient) of 0.25. Based on the data, the most correct statement is:

(A) 25% of the time the test distinguishes between ankylosing spondylitis and the normal population.
(B) 75% of the time the test distinguishes between ankylosing spondylitis and the normal population.
(C) Two-thirds of the time the test distinguishes between ankylosing spondylitis and the normal population.
(D) No conclusions can be drawn unless the p value is available.
(E) The coefficient of determination (r^2) is too low (6.25%) to make a statement.

B.   Linear Regression Analysis

9.4 Regression Analysis

Although a *scatter diagram* visually demonstrates the general path of points that illustrate the association between two variables, a **linear relationship** may be established through a procedure known as **linear regression analysis.**

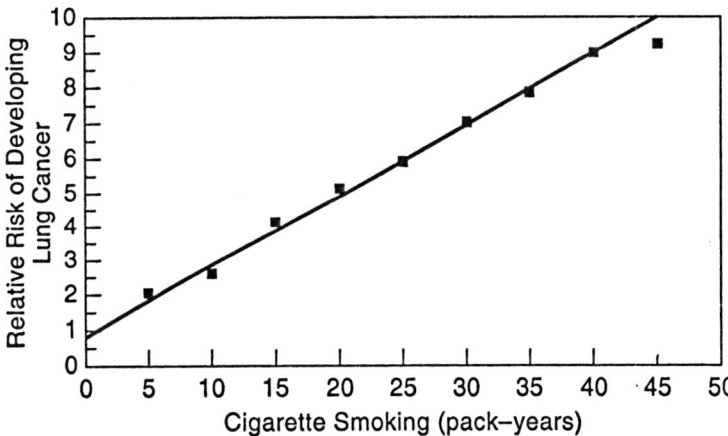

Fig. 9.3 Association Between Smoking and the Relative Risk of Lung Cancer.

By using the *least squares method* (a procedure that minimizes the vertical deviations of plotted points surrounding a straight line), we are able to construct a *"best fitting"* straight line to the scattergram points, and then formulate a **regression equation** in the form of

$$y = ax + b$$

By plotting the *independent variable* on the x axis, and the *dependent variable* on the y axis, we can use this *regression equation* to predict the value of the *dependent variable* based upon the value of the *independent variable* because all points on a straight line can be determined by substituting various values of x into the equation.

Thus, **regression analysis** permits us to **predict** *the unit change in a dependent variable for each unit change in an independent variable.* Although predictions made about the independent variable by using linear regression analysis are rarely 100% accurate, a straight line is the best way to describe the average path of scattergram points, and can be reasonably precise.

Although computations and interpretations of linear regression equations will not be tested on the examination, their *appropriateness* and their *use* in clinical medicine must be understood.

☐ NOTE

As with correlation coefficients, causation may not be assumed by using regression analysis.

E x a m p l e

As you know, CHD (coronary heart disease) has a multifactorial etiology. Suspected risk factors for acute coronary heart disease are listed in the following diagram. Each of these individual factors, either alone or in combination, has been shown to be associ-

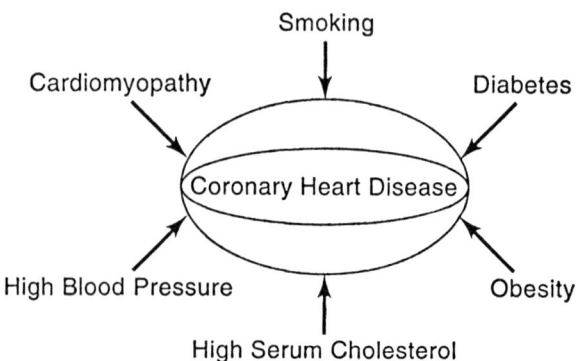

Fig. 9.4 Risk Factors for Coronary Heart Disease.

ated with this disease. *Linear regression* is used to predict the contribution of each individual factor in the etiology of acute coronary heart disease.

Summary

- **Correlation coefficient (*r*)** and **coefficient of determination (*r*^{2})** are used to measure the *degree of the association* between two variables.

- **Linear regression analysis** is used to predict one variable (dependent) based upon the value of another variable (independent) using a *regression equation.*

- **Correlation coefficient** and **regression analysis** involve the use of *scattergrams.*

- Multiple correlation coefficients, coefficients of determination, and regression analyses may be used when a dependent variable (disease or outcome) is associated with more than one independent variable (etiologic agent, risk factor, etc.).

CHAPTER 10

Practice Test

Carefully read the following instructions before taking the Practice Test.

1. This examination consists of 100 questions that are similar to those you will encounter on the actual examination. They are integrated in an effort to simulate the examination style.

2. You should allow yourself a total of 1 hour and 23 minutes for the examination. This is based on the allowance of approximately 50 seconds for each item, the approximate allowed time during the actual examination.

3. Be sure you have an adequate number of pencils and erasers, a clock, a comfortable setting, and enough distraction-free time to complete the test.

4. There is an answer sheet on page 109 for you to use in recording your answers. Although this is not the exact type of answer sheet as you will encounter on the examination, its use will allow you to assess your areas of strength and weakness in Epidemiology and Biostatistics.

Practice Test Questions

DIRECTIONS (Questions 1 through 41): Each of the numbered items or incomplete statements in this section is followed by answers. Select the one lettered answer that is best in each case.

Annual incidence rate of malignant melanoma = 30 per 1,000,000

Annual mortality rate of malignant melanoma = 5 per 1,000,000

Point prevalence rate of malignant melanoma = 90 per 1,000,000

1. With the information given, the average duration of the disease is

   (A) 3 years
   (B) 4 years
   (C) 5 years
   (D) 6 years
   (E) unable to be determined due to insufficient data

2. Three groups of subjects were followed over the course of five years to compare the efficacy of different pharmacologic treatments for sideroblastic anemia. The most appropriate statistical analysis to determine the quantitative serologic differences resulting from these treatments would be a(n)

   (A) regression analysis
   (B) *F* test (ANOVA)
   (C) correlation analysis
   (D) chi-square test
   (E) *t* test

3. In a suburb of San Francisco, it was reported that there were 300 HIV-positive people of whom 10 died in 1991. The case fatality rate per 100 people is

   (A) 0.33%
   (B) 3%
   (C) 3.3%
   (D) 3.5%
   (E) 33%

Questions 4 and 5

In a class of 134 medical students, the mean systolic blood pressure was found to be 126 mm Hg with a standard deviation of 6 mm Hg.

4. If the blood pressures in this sample are normally distributed, two-thirds of the medical students will have a systolic blood pressure in the range of (mm Hg)

   (A) 108–144
   (B) 114–138
   (C) 118–134
   (D) 120–132
   (E) 124–128

5. If the blood pressures in this sample are normally distributed, what portion of the medical students will have systolic blood pressures above 132 mm Hg?

   (A) 0.5%
   (B) 2.5%
   (C) 5%
   (D) 16%
   (E) 32%

Questions 6 through 8

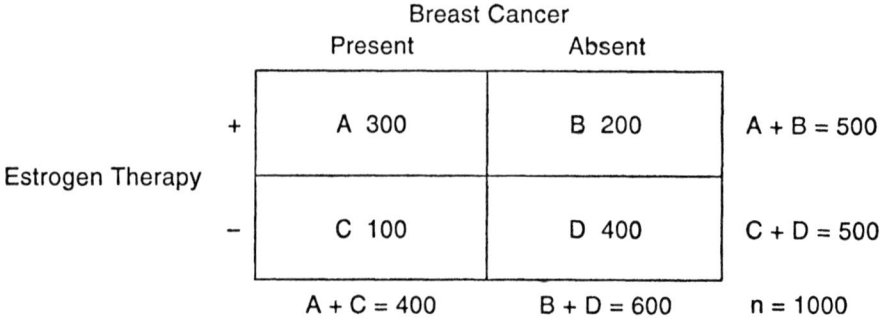

In a cohort study concerning the relationship between the use of exogenous estrogens and the subsequent risk of breast cancer, a sample of 1000 premenopausal women were followed for 8 years. The results are presented in the table above.

6. The absolute risk of breast cancer among women receiving estrogen therapy is

(A) 0.2
(B) 0.3
(C) 0.4
(D) 0.6
(E) 0.8

7. The absolute risk of breast cancer among women who did not receive estrogen therapy is

(A) 0.05
(B) 0.2
(C) 0.4
(D) 0.6
(E) 0.8

8. The relative risk associated with estrogen therapy in this study is

(A) 0.25
(B) 0.33
(C) 0.5
(D) 2
(E) 3

9. Serum cholesterol levels for two groups of Americans were recorded in 1989. The mean cholesterol level for a group of 28 African Americans was 271 mg/dL and 184 mg/dL for a group of 25 Oriental Americans. To determine whether or not these measurements were significantly different, the most appropriate statistical test would be a(n)

(A) chi-square test
(B) correlation analysis
(C) *F* test (ANOVA)
(D) regression analysis
(E) student's *t* test

10. The measure of central tendency that is most affected by extreme scores in a sample distribution is the

(A) mean
(B) median
(C) mode
(D) standard deviation
(E) variance

11. In a prospective study, the occurrence of transitional cell carcinoma of the bladder was recorded for smokers and nonsmokers. The difference in incidence between smokers and nonsmokers was reported to be significant at the $p < 0.05$ level. The most appropriate statement that can be made about this study is that

(A) the null hypothesis may be rejected even though the results could have occurred purely by chance a maximum of 5% of the time

(B) a significant difference in the incidence between smokers and nonsmokers may occur 5% of the time or less

(C) the null hypothesis may be accepted because there is a difference in the incidence rates between smokers and nonsmokers only 5% of the time

(D) the alternate hypothesis may be rejected because the null hypothesis is true up to 5% of the time

(E) a causal relationship between smoking and the incidence of transitional-cell carcinoma of the bladder may be established at a significance level of 0.05

Questions 12 and 13

To study the association between coffee drinking and cancer of the pancreas, the coffee consumption (ounces per month) for two groups of postal workers was recorded. The results were as follows:

	MEAN	STANDARD DEVIATION
Group 1	146 oz.	21 oz.
Group 2	238 oz.	24 oz.

t value = 2.77
Degree of freedom (d_f) = 20

STUDENT'S t DISTRIBUTION TABLE

DEGREE OF FREEDOM	PROBABILITY (P VALUE)			
	0.50	0.10	0.05	0.01
1	1.000	6.31	12.71	63.66
5	0.727	2.02	2.57	4.03
10	0.700	1.81	2.23	3.17
20	0.687	1.71	2.06	2.79
25	0.684	1.71	2.06	2.79
Infinity	0.674	1.64	1.96	2.58

12. The difference in coffee consumption between these two groups has a significance level of

(A) $p < 0.50$
(B) $p < 0.10$
(C) $p < 0.05$
(D) $p < 0.01$
(E) $p < 0.001$

13. If the calculated t value in this study had been 2.79, and all other values remained the same, the level of significance would then be

(A) less
(B) unknown
(C) greater
(D) unchanged
(E) negligible

14. In a study measuring depression immediately following a spontaneous abortion, 1500 women volunteered to be interviewed in their home shortly after the occurrence of this experience. One of the problems encountered in this household health survey was noncompliance with the appointments scheduled at their homes by the researchers. The most acceptable way to manage the problem of nonresponse is to

(A) send a self-addressed stamped envelope with a written questionnaire for the nonresponsive volunteers to fill out and mail to the survey center

(B) arrange an interview with the husband or close relative who knows the nonresponding volunteer well enough to answer the survey questions correctly

(C) record responses from one of the questionnaires that is considered to contain typical answers as a substitute for the answers of the nonresponders

(D) record the survey answers of the nonresponders as unknown in the data analysis if repeated visits to their home are unsuccessful

(E) completely omit the nonresponders from the survey analysis after repeated visits to their home are unsuccessful

15. The case fatality rate for disease X is 20% within three years of the initial diagnosis. The probability that three randomly selected patients with this disease will die within the same period is

(A) 0.8
(B) 0.08
(C) 0.008
(D) 0.0008
(E) 0.00008

16. The power of a statistical test can be determined by using the formula

(A) $1 - H_0$
(B) $1 - H_A$
(C) $1 - \alpha$
(D) $1 - \beta$
(E) $H_0 - H_A$

Questions 17 and 18

A sample of 175 nursing students was found to have a mean serum cholesterol level of 186 mg/dL with a variance of 36 mg/dL.

17. If the data is normally distributed, one-third of the nursing students will have a cholesterol level (mg/dL) outside the range of

(A) 150–222
(B) 168–204
(C) 174–198
(D) 180–192
(E) 186–192

18. If the data is normally distributed, the cholesterol levels of 95% of the nursing students will be in the range (mg/dL) of

(A) 150–222
(B) 168–204
(C) 174–198
(D) 180–192
(E) 186–192

19. While investigating the association between elevated levels of alpha feto protein and the development of neural tube defects, maternal histories of 3250 anencephalic fetuses were studied. If a maximum intercorrelation (r) of 0.81 is found to exist with respect to elevated levels of alpha feto protein in pregnant women and the subsequent consequence of fetal anencephaly, which of the following statements is correct?

(A) 81% of the time, anencephaly is associated with elevated maternal levels of alpha feto protein.
(B) 75% of the time, anencephaly is associated with elevated maternal levels of alpha feto protein.
(C) Two-thirds of the time, anencephaly is associated with elevated maternal levels of alpha feto protein.
(D) No conclusions can be drawn unless the p value is known.
(E) The maximum intercorrelation is too low (6.25%) to establish a true association between alpha feto protein levels and the occurrence of anencephaly.

20. In a case-control study, the distortion of risk ratios that occurs as a result of different risk factor probabilities among cases and controls is referred to as

(A) Berkson's bias
(B) observer bias
(C) interviewer bias
(D) unacceptability bias
(E) confounding

21. A typical or average value found in a set of data is referred to as a measure of

(A) range
(B) standard deviation
(C) standard error of the mean
(D) variance
(E) central tendency

22. The time interval between remission and the reappearance of symptoms is referred to as the

(A) incubation period
(B) latent period
(C) recovery period
(D) infectivity period
(E) communicable period

23. Following a large group of cigarette smokers for a period of 10 years to determine the occurrence of chronic obstructive pulmonary disease (COPD), coronary heart disease, and various forms of lung cancer would be an example of a

(A) randomized clinical trial
(B) cross-sectional study
(C) prevalence study
(D) cohort study
(E) case-control study

24. A test that measures what it was originally designed to measure is said to have

(A) sensitivity
(B) specificity
(C) validity
(D) reliability
(E) predictability

25. A nutritional research team followed serum levels of vitamin B_{12} and folic acid in 125 children for five years to determine the association between cyanocobalamin deficiency and the subsequent risk of developing megaloblastic anemia. The results were as follows:

VITAMIN B_{12} LEVELS	
mean	262 pg/mL
median	228 pg/mL
mode	196 pg/mL

From the data, it can be concluded that this distribution is

(A) normal
(B) positively skewed
(C) negatively skewed
(D) skewed toward the left
(E) unable to be identified

26. Drugs X, Y, and Z were given daily to HIV-positive subjects at

  i. low dosage
  ii. moderate dosage
  iii. high dosage

T-cell counts were then conducted every three months for a period of five years. The most appropriate statistical analysis for determining the significance of the differences between the proportions of T-suppressor and T-helper cells resulting from these dosages would be a(n)

(A) student's t test
(B) F test (ANOVA)
(C) chi-square test
(D) correlation analysis
(E) regression analysis

Questions 27 and 28

The five-year survival rates of patients with aplastic anemia secondary to hypernephroma are represented by the following figure:

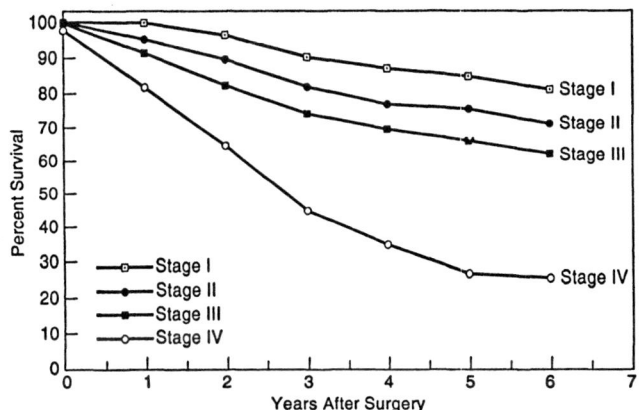

Survival of Patients with Hypernephroma.

27. The two-year survival of a patient with Stage II hypernephroma in this study is

(A) 30%
(B) 40%
(C) 50%
(D) 60%
(E) 90%

28. In this study, the median survival for patients with Stage IV hypernephroma was

(A) 2.5 years
(B) 3.5 years
(C) 4.5 years
(D) 5.5 years
(E) 6.5 years

29. In a prospective study comparing the effectiveness of two chemotherapeutic treatments for cervical cancer, cases were selected from one group of patients who had annual Pap smears for at least five years prior to their initial diagnosis and from another group of patients who had no history of prior Pap smear screening. The selection of cases from both groups in this study may result in

(A) confounding
(B) interviewer bias
(C) Berkson's bias
(D) lead time bias
(E) recall bias

30. The incidence and prevalence rates of Disease *X* found in women between the ages of 35 and 65 over a ten-year period is illustrated in the following figure:

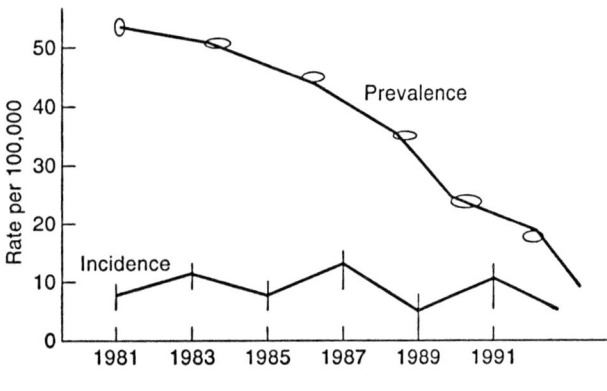

Women with Disease *X*.

All of the following explanations about the data are correct except:

(A) The prevalence of the disease is decreasing.
(B) The duration of the disease is decreasing.
(C) Recovery from the disease is becoming rapid.
(D) The prevalence of the disease is inversely proportional to its incidence.
(E) The incidence is relatively constant throughout the duration of the study.

31. The Gaussian distribution curve characteristics include

(A) the total area under the curve represents all possible values
(B) the mean, median, and mode are found below the apex of the curve
(C) one-third of the values are found more than one standard deviation from the mean
(D) 2.5% of the values are found two standard deviations above the mean
(E) All the above

CHI-SQUARE (χ^2) DISTRIBUTION TABLE

DEGREE OF FREEDOM	PROBABILITY (*P* value)				
	0.99	0.95	0.05	0.01	0.001
1	0.000157	0.00393	3.841	6.63	10.83
2	0.0201	0.103	5.991	9.21	13.82
3	0.115	0.352	7.815	11.34	16.27
4	0.297	0.711	9.488	13.28	18.47
5	0.554	1.145	11.070	15.09	20.51

32. To assess the association between a deficiency of the enzyme hexosaminidase A and the subsequent risk of developing Tay-Sachs disease, several groups of premature infants were investigated. The chi-square value for this study was reported to be 13.22 with 4 degrees of freedom. From the chi-square distribution table, the significance level of this study is

(A) $p < 0.99$
(B) $p < 0.95$
(C) $p < 0.05$
(D) $p < 0.01$
(E) $p < 0.001$

33. If, in one of the groups of premature infants, the maximum value for hexosaminidase A was substitued with a much higher value, all of the following measures in that group may be affected by that score, except the

(A) variance
(B) range

(C) standard deviation
(D) median
(E) mean

34. A distribution in which the median is greater than the mean is described as being

(A) skewed toward the left
(B) skewed toward the right
(C) positively skewed
(D) only slightly skewed
(E) moderately skewed

35. In a case-control study concerning the relationship between phenylbutazone and the subsequent risk of neutropenia, medical records of 300 children were investigated. The results were as follows:

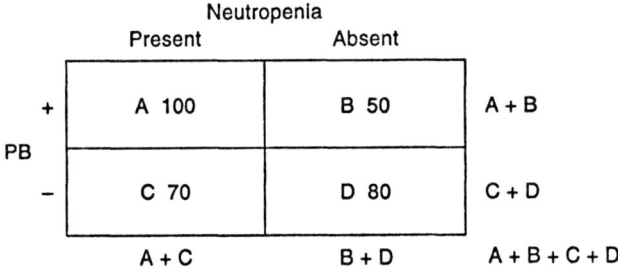

PB +   Confirmed history of phenylbutazone treatment
PB −   No history of phenylbutazone treatment

The odds ratio in this study is
(A) $(100 \times 80)/(100 \times 70)$
(B) $(100 \times 80)/(100 \times 50)$
(C) $(100 \times 80)/(50 \times 70)$
(D) $(100 \times 50)/(70 \times 80)$
(E) $(100 \times 70)/(50 \times 80)$

36. If 20% of a given population suffers from hypertension and 10% from hypotension, the probability that a randomly selected member of this population will suffer from one of these ailments is

(A) 15%
(B) 20%

(C) 25%
(D) 30%
(E) 35%

Questions 37 and 38

In a longitudinal study of the relationship between paroxysmal nocturnal hemoglobinuria and the subsequent onset of complement-sensitive associated anemia, 150 children were followed for a period of three years.

37. At a significance level of $p < 0.01$, the null hypothesis states that

(A) when compared to normal children, there is a difference in the incidence of complement-sensitive associated anemia in children with a history of paroxysmal nocturnal hemoglobinuria
(B) when compared to normal children, there is no difference in the incidence of complement-sensitive associated anemia in children with a history of paroxysmal nocturnal hemoglobinuria
(C) when compared to normal children, there is a difference in the incidence of complement-sensitive associated anemia in children with a history of paroxysmal nocturnal hemoglobinuria 1% of the time
(D) when compared to normal children, there is no difference in the incidence of complement-sensitive associated anemia in children with a history of paroxysmal nocturnal hemoglobinuria 1% of the time
(E) when compared to normal children, there is no significant difference in the incidence of complement-sensitive associated anemia in children with a history of paroxysmal nocturnal hemoglobinuria 99% of the time

38. At a significance level of $p < 0.05$, the alternate hypothesis states that

(A) there is an insignificant difference between children with a history of paroxysmal nocturnal hemoglobinuria and normal children with respect to the incidence of complement-sensitive associated anemia

(B) there is no difference between children with a history of paroxysmal nocturnal hemoglobinuria and normal children with respect to the incidence of complement-sensitive associated anemia

(C) there is a difference between children with a history of paroxysmal nocturnal hemoglobinuria and normal children with respect to the incidence of complement-sensitive associated anemia 5% of the time

(D) there is a difference between children with a history of paroxysmal nocturnal hemoglobinuria and normal children with respect to the incidence of complement-sensitive associated anemia 95% of the time

(E) there is no difference between children with a history of paroxysmal nocturnal hemoglobinuria and normal children with respect to the incidence of complement-sensitive associated anemia 95% of the time

39. In a cohort study involving the relationship between HIV status and the subsequent risk of developing pneumocystis carinii pneumonia, 50 HIV-positive volunteers were followed for 6 months: 100 for 1 year, 100 for 3 years, and 200 for 5 years. The number of person-years of observation in this study was

(A) 9.5
(B) 1425
(C) unable to be determined for different periods
(D) unable to be determined without a mortality rate
(E) unable to be determined without an incidence rate

40. The persistent presence of benign prostatic hyperplasia in a large metropolitan area represents a(n)

(A) endemic
(B) hyperendemic
(C) endemoepidemic
(D) epidemic
(E) pandemic

41. A Pap smear and colposcopic examination for the early detection of cervical cancer and papilloma virus infection constitute

(A) primary prevention
(B) secondary prevention
(C) tertiary prevention
(D) medical treatment
(E) surgical treatment

DIRECTIONS (Questions 42 through 100): Each of the numbered items or incomplete statements in this section is followed by a series of lettered answers. For each item, select the one lettered answer that is most closely associated with it. Each lettered item may be selected once, more than once, or not at all.

Questions 42 through 45

A new radiologic scanning test for the early detection of osteogenic sarcoma was used to evaluate 500 patients. The results of the study were as follows:

	OSTEOGENIC SARCOMA		
	PRESENT	ABSENT	
Test +	100	150	(100 + 150)
Test −	75	175	(75 + 175)
	(100 + 75)	(150 + 175)	$n = 500$

Match the following:

(A) $175/(150 + 175)$
(B) $100/(100 + 150)$
(C) $150/(150 + 175)$
(D) $150/(100 + 150)$
(E) $100/(100 + 75)$
(F) $175/(75 + 175)$

42. Negative predictive value

43. Sensitivity

44. Positive predictive value

45. Specificity

Questions 46 through 49

In a cohort study concerning the relationship between cirrhosis of the liver and the subsequent risk of hepatic encephalopathy, 1000 adults were followed from 1986 to 1990. The results were as follows:

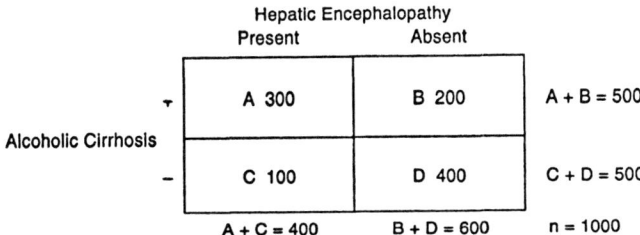

46. The absolute risk for the onset of hepatic encephalopathy among adults with cirrhosis of the liver is

(A) 30%
(B) 50%
(C) 60%
(D) 67%
(E) 75%

47. The relative risk in this study is

(A) 2.5
(B) 3.0
(C) 3.75
(D) 5.0
(E) 6.0

48. The attributable risk in this study is

(A) 0.25
(B) 0.40
(C) 0.50
(D) 0.67
(E) 0.75

49. The attributable risk percent in this study is

(A) 30%
(B) 50%
(C) 60%
(D) 67%
(E) 75%

Questions 50 through 55

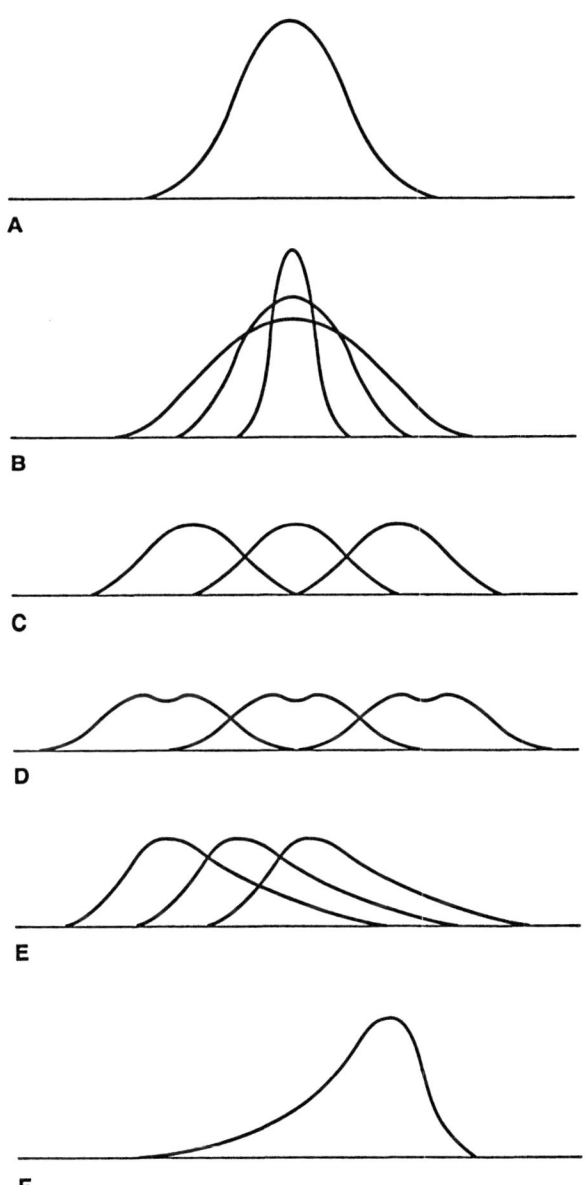

Match the lettered illustrations with the following distributions:

50. Negatively skewed

51. Equal variances with different means

52. Normal

53. Positively skewed

54. Bimodal

55. Equal means with different variances

Questions 56 through 59

Match the following with the numbered questions:

   (A) Type I error
   (B) Type II error
   (C) Probability error
   (D) Correct decision
   (E) Irrelevant decision

56. Rejecting the null hypothesis when H_0 is false

57. Accepting the null hypothesis when H_0 is false

58. Rejecting the null hypothesis when H_0 is true

59. Accepting the null hypothesis when H_0 is true

Questions 60 through 73

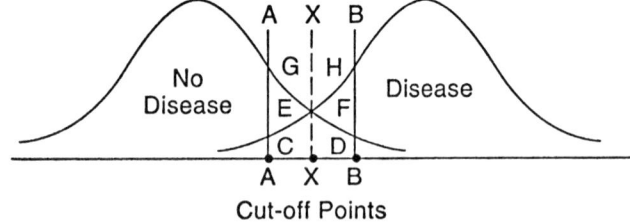

Diagnostic Screening Test.

   (A) A   (E) E
   (B) B   (F) F
   (C) C   (G) G
   (D) D   (H) H

With X representing the most accurate cutoff point for a diagnostic screening test, match the following with the most appropriate letter in the figure:

60. Cutoff point of lesser specificity

61. True-positives and false-positives for cut-off point X

62. Cutoff point of greater false-positive rate

63. True negatives for cutoff point X

64. False negatives for cutoff point B

65. Cutoff point set too high

66. Cutoff point of greater sensitivity

67. False positives for cutoff point A

68. True positives for cutoff point X

69. Cutoff point of greater specificity

70. Cutoff point of greater false-negative rate

71. Cutoff point set too low

72. Cutoff point of lesser sensitivity

73. True negatives and false negatives for cutoff point X

Questions 74 through 77

Refer to this table for the following questions:

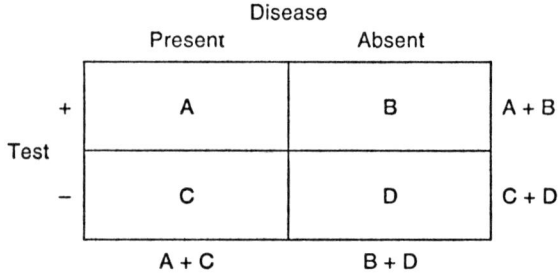

74. False positives appear in cell

(A) A
(B) B
(C) C
(D) D
(E) E

75. False negatives appear in cell

(A) A
(B) B
(C) C
(D) D
(E) E

76. True positives appear in cell

(A) A
(B) B
(C) C
(D) D
(E) E

77. True negatives appear in cell

(A) A
(B) B
(C) C
(D) D
(E) E

Questions 78 through 84

To evaluate a new noninvasive radiologic scanning procedure for myelofibrosis, a group of 200 patients volunteered to be screened. Among this group, 100 were known to have the disease as confirmed by medical records. During the study, 130 were found to be positive for the test, of which 60 were false positives, and 30 of the 100 patients who were known to have the disease were false negatives. The following ratios were evaluated after the screening:

(A) $60/(60 + 40) = 0.60$
(B) $40/(60 + 40) = 0.40$
(C) $30/(70 + 30) = 0.30$
(D) $40/(30 + 40) = 0.57$
(E) $70/(30 + 70) = 0.70$
(F) $70/(70 + 60) = 0.54$
(G) $(70 + 40)/(70 + 60 + 30 + 40) = 0.55$

Match each of the following measurements with its corresponding ratio:

78. Sensitivity

79. Specificity

80. Positive predictive value

81. Negative predictive value

82. False-positive rate

83. False-negative rate

84. Accuracy of the test

Questions 85 and 86

(A) Standard error
(B) Confounding error
(C) Systemic error
(D) Random error
(E) Probability error

Match the following with the best answer:

85. Reduced precision in a test

86. Reduced accuracy in a test

Questions 87 through 90

Match the scattergrams shown below:

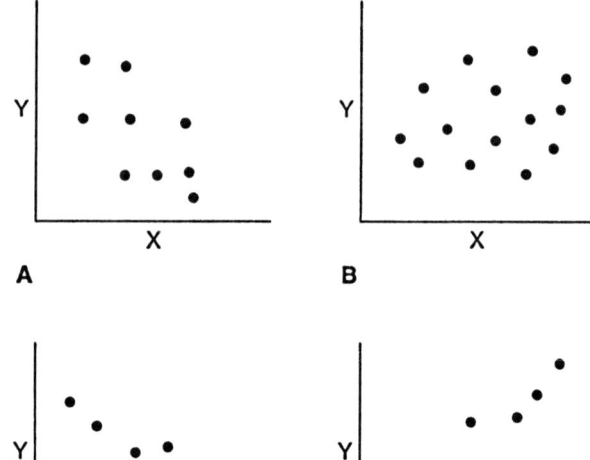

A   B

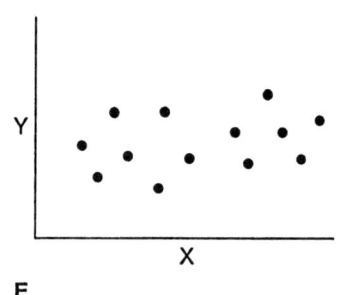

C   D

E

87. High positive correlation

88. Low positive correlation

89. Correlation of zero

90. High negative correlation

Questions 91 through 94

(A) Dependablity of a test
(B) Consistency and reproducibility of a test
(C) Absence of random variability in a test
(D) Trueness of test measurements
(E) Appropriateness of a test
(F) B and C

91. Validity

92. Reliability

93. Precision

94. Accuracy

Questions 95 through 99

	Truth	
	H_0 True	H_0 False
Accept H_0	A	B
Reject H_0	C	D

Decision

(A) A
(B) B
(C) C
(D) D
(E) A and D
(F) B and C
(G) None of the above

Match each of the following questions with the correct cell choice:

95. Correct decision

96. Type I error

97. Type II error

98. Used to determine the power of a test

99. The investigator has some control of this error

100. Statistical methods are used in medical research because they

(A) provide values for results that are dependent upon chance

(B) provide conclusions about cause and effect

(C) provide controls for variables that were absent in the study design

(D) provide assurance of the significance of the findings

(E) provide controls for some of the more common sources of experimental error

EPIDEMIOLOGY
&
BIOSTATISTICS

ANSWERS AND EXPLANATIONS

ANSWERS AND EXPLANATIONS

1. **(A).** The relationship between the incidence, prevalence, and duration of a disease is expressed by the formula: prevalence = incidence × duration of the disease; 900 = 300 × duration of the disease. Therefore, the duration of the disease is 3 years. For further explanation, refer to page 28.

2. **(B).** The F test (ANOVA) is used to compare the means of three or more samples or groups (as opposed to the student's t test, which compares the means of two samples or groups) for the purpose of determining the statistical significance of an observed finding. For further explanation, refer to page 59.

3. **(C).** Case fatality rate by definition is:

$$\frac{\text{Number of deaths due to a disease}}{\text{Number of people with the same disease}} \times 10^x$$

Usually expressed as a percentage: $(10/300) \times 100 = 3.3\%$

For further explanation, refer to page 30.

4. **(D).** 68% or two-thirds of the values fall within one standard deviation of the mean. 126 mm Hg plus or minus 6 mm Hg results in a range of 120 to 132 mm Hg. For further explanation, refer to page 52.

5. **(D).** Because 132 mm Hg is exactly one standard deviation above the mean, it represents the upper limit of the range of systolic blood pressures within which 68% of the values will fall—that is, between 120 and 132 mm Hg (see previous question). Of the remaining 32% of the values, half of them (16%) will fall below 120 mm Hg and half will lie above 132 mm Hg. For further explanation, refer to page 52.

6. **(D).** The absolute risk for breast cancer among women who received estrogen therapy is calculated by the formula:

$$\frac{A}{(A + B)} = \frac{300}{(300 + 200)} = 0.6$$

For further explanation, refer to page 45.

7. **(B).** The absolute risk for breast cancer among women who did not receive estrogen therapy is calculated by the formula:

$$\frac{C}{(C + D)} = \frac{100}{(100 + 400)} = 0.2$$

For further explanation, refer to page 45.

8. **(E).** The relative risk is calculated by the formula:

$$\frac{A/(A + B)}{C/(C + D)} = \frac{300/(300 + 200)}{100/(100 + 400)} = \frac{3/5}{1/5} = 3$$

For further explanation, refer to page 45.

9. **(E).** The t test is used to compare the means of two small ($n < 30$) samples or groups (as opposed to the F test, which compares the means of three or more samples or groups) for the purpose of determining the statistical significance of

an observed finding. For further explanation, refer to page 57.

10. **(A).** The mean is the measure of central tendency that is most affected by extreme scores. Neither the standard deviation nor the variance is a measure of central tendency. For further explanation, refer to page 49.

11. **(A).** A significance level of $p < 0.05$ states that the results of the study could have occurred purely by chance only 5% of the time or less. This is the probability (p value) that represents the lowest significance level at which the null hypothesis (H_0) may be rejected in this particular study. Most researchers use $p <0.05$ to reject the null hypothesis (H_0), which is fairly arbitrary but universally accepted. For further explanation, refer to page 22.

12. **(C).** After locating the row in the t distribution table corresponding to 20 degrees of freedom (d_f), select the t value in that row that most closely approximates, but does not exceed, the given value of 2.77. As can be seen from the table, 2.06 is the highest critical value that is less than or equal to ($\leq$) 2.77, and its location is in the column corresponding to a p value of < 0.05. For further explanation, refer to page 58.

> **WARNING:** The comprehensive medical examinations may not label the columns and rows of the standard t distribution table. You must know that the first column on the left is for degrees of freedom (d_f), the top row is for p values (levels of significance), and the remainder of the table is for t values (critical values).

13. **(C).** Using the same method as in the previous question, it can be seen from the t distribution table that with 20 degrees of freedom, the critical value (t score) of 2.79 lies in the column corresponding to a p value of < 0.01. This represents a level of significance greater than the previous level of < 0.05 because the probability of the results occurring purely by chance is now only 1% (0.01) as opposed to the previous probability of 5% (0.05). For further explanation, refer to page 58.

14. **(D).** To reduce the potential for selection (sampling) bias, nonresponders should be included in a survey, with their results acknowledged as unknown in the data analysis. For further explanation, refer to page 44.

15. **(C).** To calculate the probability (P) for the combined occurrence of three independent events, the multiplication rule is used. Events are said to be independent if the occurrence of one event has no effect upon the occurrence of the other event. The probability that all three independent events will occur is calculated by the formula

$$P(A, B, \text{and } C) = P(A) \times P(B) \times P(C)$$
$$= 0.2 \times 0.2 \times 0.2 = 0.008$$

Therefore, the probability that disease X will result in the fatality of all three randomly selected cases within three years of their initial diagnosis is 0.008. For further explanation, refer to page 24.

16. **(D).** The power of a test is determined by using the formula $1 - \beta$ (beta or type II error). For further explanation, refer to page 22.

17. **(D).** With a variance of 36, the standard deviation (square root of the variance) is conveniently seen to be 6 mg/dL. In a normally distributed sample population, we know that 68% (two-thirds) of the values fall within one standard deviation of the mean and one-third of the values do not. Therefore, one-third of the nursing students will have serum cholesterol levels that lie outside the range of 186 plus or minus 6 = 180 to 192 mg/dL. For further explanation, refer to page 52.

18. **(C).** 95% of the values fall within two standard deviations of the mean. Therefore, the range of values will be 186 plus or minus 12 = 174 to 198 mm Hg. For further explanation, refer to page 52.

19. **(C).** A correlation coefficient (r) of 0.81 yields a coefficient of determination (r^2) equal to 0.66. Therefore, 66% (2/3rds) of the variation (anencephaly) may be directly associated with elevated maternal levels of alpha feto protein. For further explanation, refer to page 68.

20. **(A).** This type of study (retrospective, or case-control study) is prone to a form of selection (sampling) error known as Berkson's bias. This nonrandom (systemic) error involves distortions in risk ratios that are caused by the different risk factor probabilities that exist among cases and controls that are the result of different hospital admission rates for each risk-factor group. None of the remaining biases involve this source of error. For further explanation, refer to page 44.

21. **(E).** A measure that describes a typical or average value in a set of data is referred to as measure of central tendency. The standard deviation, variance, and standard error of the mean are measures of dispersion, and the range represents the spread between the highest and lowest scores in a sample. For further explanation, refer to page 49.

22. **(B).** A latent period is defined as the period of subclinical infection during which an active pathogen becomes dormant. This period begins with the remission of signs or symptoms and ends when the virus becomes reactivated and signs or symptoms reappear. For further explanation, refer to page 3.

23. **(D).** In a cohort or prospective study, a group of individuals are followed over a specified period to determine how many develop a particular characteristic or disease after the exposure to a risk factor or other independent variable. For further explanation, refer to page 35.

24. **(C).** Validity is the degree to which a test measures what it was originally designed to measure. Sensitivity and specificity (two components of validity) are the degrees to which a test correctly identifies the presence and absence of a disease respectively. The predictability of positive and negative test results is estimated by a test's positive predictive value and negative predictive value and is influenced by disease prevalence. Reliability is the degree to which a test is considered to be dependable. A test's consistency and reproducibility are dependent upon its precision, or absence of random variability. For further explanation, refer to pages 11–12.

25. **(B).** When the mean is greater than the median, a distribution is said to be positively skewed or skewed toward the right (with respect to the x axis). An easy way to avoid confusion regarding skewed distributions is to remember that from the tail of the curve to the apex (mode), the mean, median, and mode are always in alphabetical order, and the mean always follows the tail of a skewed curve. If the tail is on the right, the mean is to the right (greater than) the median, and the curve is positively skewed. The reverse is true when the tail of the curve is on the left. For further explanation, refer to page 55.

26. **(C).** The chi-square test is used for the comparison of two or more independent proportions within two or more sample populations. For further explanation, refer to page 60.

27. **(E).** From the two-year survival point on the x axis, draw a straight line to intersect the Stage II curve and then connect this point with another straight line to the y axis, which corresponds to a 90% survival rate. For further explanation, refer to page 50.

28. (A). This is just the opposite of what was done in the previous question. From the median (50%) survival point on the y axis, draw a straignt line to intersect the Stage IV curve and then connect this point with another straight line to the x axis, which corresponds to 2.5 years after surgery as the length of time that half of those patients survived. For further explanation, refer to page 50.

29. (D). The two groups that were involved in this study may have differed from each other with respect to the period between diagnosis and treatment. The first group probably entered the treatment program much earlier in the course of their disease as a result of annual Pap smears, as opposed to the second group who, without Pap smear screening, may have entered the treatment program later in the course of their disease. When compared to the second group, a more favorable response to treatment by the first group may be attributable to earlier detection and not necessarily to superior treatment. The selection of cases from both of these groups introduces a form of nonrandom (systemic) error known as lead time bias. For further explanation, refer to page 44.

30. (D). As can be seen in the question illustration, the prevalence of this disease had clearly been decreasing during this ten-year period, while the incidence remained fairly constant. In light of these two findings and the knowledge that prevalence is the product of the incidence times the duration of the disease, the only circumstance under which the prevalence of a disease can fall while the incidence remains constant is if the duration of the disease decreases—that is, the recovery from the disease becomes more rapid. There is no inverse proportionality between incidence and prevalence. For further explanation, refer to page 28.

31. (E). The total area under a Gaussian distribution curve represents all possible values in the normal population. Because the mean, median, and mode are equal, they are all found below the highest point of the curve (apex) at the position of the score with the highest frequency, the mode. Because two-thirds of the values in a normal distribution are found within one standard deviation of the mean, one-third will be found beyond one standard deviation from the mean. Because roughly 95% of the values in this distribution lie within two standard deviations of the mean, 5% of the values will fall outside this region: 2.5% above and 2.5% below. For further explanation, refer to page 52.

32. (C). Reading the row in the given chi-square table (from left to right) corresponding to 4 degrees of freedom, the critical value that is less than or equal to the chi-square value given in the study (13.22) is 9.488. The location of this chi-square value is in the column represented by a significance level (p value) of < 0.05. This represents the lowest level at which the null hypothesis may be rejected in this study. Therefore, at the $p < 0.05$ level, there is a significant difference in the incidence of Tay-Sachs disease in premature infants with a deficiency of hexosaminidase A when compared to those premature infants without this deficiency. For further explanation, refer to page 60.

33. (D). When an extremely high score is substituted for another score in a sample, the mean, standard deviation, and variance may all be affected. Raising the maximum score increases the upper limit of the range in a sample. The substitution of the highest score by one that is higher will not have an effect upon the middle score, the median. For further explanation, refer to page 49.

34. (A). In a distribution where the mean is less than the median, the distribution is said to be negatively skewed or skewed toward the left. Since the mean always follows the tail of the curve and all measures of central tendency lie in alphabetical order (from tail to apex) in skewed distributions, the mean (with its tail) can lie only toward the left on the x axis if its score is less than the median (and mode). The curve, therefore, is skewed in that direction. For further explanation, refer to page 55.

35. **(C).**   The formula for odds ratio is:

$$\text{Odds Ratio} = \frac{A \times D}{B \times C} = \frac{100 \times 80}{50 \times 70}$$

For further explanation, refer to page 39.

36. **(D).**   Because hypertension and hypotension are considered to be mutually exclusive events—that is, a subject may fall into only one of these two categories—the probability that a member of this population will have one of these two ailments is calculated by using the addition rule.

$$P(A \text{ or } B) = P(A) + P(B) = 20\% + 10\% = 30\%$$

For further explanation, refer to page 24.

37. **(B).**   The null hypothesis (H_0) states that there is no difference between children with a history of paroxysmal nocturnal hemoglobinuria and children without this history with respect to the development of complement-sensitive associated anemia. The observed difference, if any, is by chance alone. For further explanation, refer to page 21.

38. **(D).**   At a significance level of $p < 0.05$ the alternate hypothesis (H_A) states that 95% of the time there is a difference between children with a history of paroxysmal nocturnal hemoglobinuria and children without this history with respect to the incidence of complement-sensitive associated anemia. Any differences that occurred purely by chance (H_0) could have occurred only 5% of the time or less. For further explanation, refer to page 21.

39. **(B).**   Person-years is defined as the number of persons followed in a study multiplied by the number of years of observation. The number of person-years of observation involved in this study can be calculated as follows:

50 people × .5 years =	25 person-years
100 people × 1 year =	100 person-years
100 people × 3 years =	300 person-years
200 people × 5 years =	1000 person-years
Total	= 1425 person-years

Incidence and mortality rates are not necessary for this calculation. For further explanation, refer to page 29.

40. **(A).**   Endemics are diseases, conditions, or other health-related events that are constantly present in a particular community. The persistent presence of benign prostatic hyperplasia in a large metropolitan area falls into this category.

Epidemics are diseases, conditions, or other health-related events that typically have only an occasional presence but whose incidence in a given population has increased to a level that is clearly greater than was anticipated.

Pandemics are geographically widespread epidemics, and endemoepidemics are endemics that occasionally become epidemic. Hyperendemics are diseases that are constantly present, have a high incidence, and have an effect upon all age groups in a particular population. For further explanation, refer to page 1.

41. **(B).**   Secondary prevention involves the early detection and treatment of disease. Tertiary prevention follows secondary prevention with treatment designed to prevent complications secondary to the illness and to improve the overall function of the patient. Primary prevention, by contrast, is initiated prior to the onset of disease and is intended to completely circumvent its occurrence. Medical and surgical treatment are two components of secondary prevention. For further explanation, refer to page 2.

42–45. **42-(F), 43-(E), 44-(B), and 45-(A).**

		OSTEOGENIC SARCOMA		
		PRESENT	ABSENT	
Test	+	A 100	B 150	(100 + 150) = A + B
	−	C 75	D 175	(75 + 175) = C + D
		(100 + 75)	(150 + 175)	n = 500
		A + C	B + D	

42. Negative predictive value

$$= \frac{D}{C+D} = \frac{175}{75+175}$$

43. Sensitivity $= \dfrac{A}{A+C} = \dfrac{100}{100+75}$

44. Positive predictive value

$$= \frac{A}{A+B} = \frac{100}{100+150}$$

45. Specificity $= \dfrac{D}{B+D} = \dfrac{175}{150+175}$

For further explanation, refer to page 16.

46–49. **46-(C), 47-(B), 48-(B), and 49-(D).**

		Hepatic Encephalopathy		
		Present	Absent	
	+	A 300	B 200	A + B = 500
Alcoholic Cirrhosis				
	−	C 100	D 400	C + D = 500
		A + C = 400	B + D = 600	n = 1000

46. Absolute risk $= \dfrac{A}{A+B} = \dfrac{300}{300+200} = 60\%$

47. Relative risk

$$= \frac{A/(A+B)}{C/(C+D)} = \frac{300/(300+200)}{100/(100+400)} = \frac{3/5}{1/5} = 3$$

48. Attributable risk $= A/(A + B) - C/(C + D)$

$$= \frac{300}{300+200} - \frac{100}{100+400} = 0.6 - 0.2 = 0.4$$

49. Attributable risk percent

$$= \frac{\text{Attributable risk}}{\text{Absolute risk}} = \frac{0.4}{0.6} = 0.666 \times 100 = 67\%$$

For further explanation, refer to page 47.

50. **(F).** Negatively skewed distributions are those that are skewed toward the left —that is, the direction of the tail is toward the left (negatively directed) along the *x* axis. For further explanation, refer to page 55.

51. **(C).** The symmetrical (bell-shaped) curves in this figure all have equal widths, illustrating equivalent standard deviations and variances, but they differ with respect to the positions of their apices (modes) along the *x* axis. Because the mean, median, and mode are equal in a bell-shaped (normal) distribution, all three measures of central tendency are different in the three figures. For further explanation, refer to page 56.

52. **(A).** This figure represents the classic bell-shaped normal (Gaussian) distribution. For further explanation, refer to page 56.

53. **(E).** Positively skewed distributions are those that are skewed toward the right —that is, the direction of the tail is toward the right (positively directed) along the *x* axis. For further explanation, refer to page 56.

54. **(D).** A bimodal distribution can be identified by the presence of two humps (modes) of equal or unequal size. For further explanation, refer to page 56.

55. **(B).** The symmetrical (bell-shaped) curves in this illustration have a common midline representing the mean, but the spread from this midline (standard deviation, and therefore variance) is different in each curve. For further explanation, refer to page 56.

56–59. **56-(D), 57-(B), 58-(A), and 59-(D).** Rejecting the null hypothesis (H_0) when it is true is a Type I error, and accepting the null hypothesis when it is false is a Type

II error. Accepting and rejecting H_0 when it is true and false, respectively, are correct decisions. Although probability miscalculations certainly introduce error, they are not specific for Type I or Type II errors. None of the decisions discussed in this question may be considered irrelevant. For further explanation, refer to page 22.

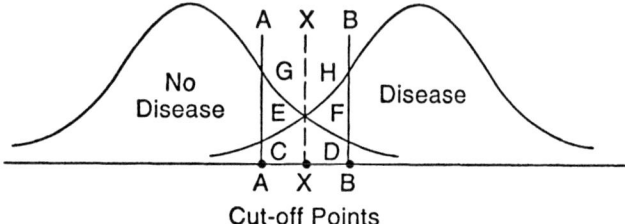

Cut-off Points

Diagnostic Screen Test

60–73. 60-(A), 61-(D), 62-(A), 63-(E), 64-(F), 65-(B), 66-(A), 67-(E), 68-(F), 69-(B), 70-(B), 71-(A), 72-(B), and 73-(C). The cutoff point of a diagnostic screening test has an effect on its sensitivity and specificity. Point X in the illustration is the most desirable of the three cutoff points from a statistical point of view (not necessarily from a clinical perspective) because it is at this point where the sensitivity and specificity of the test results are most equally balanced. If the cutoff point of a test is set lower (point A in the figure), the results will show greater sensitivity than with cutoff point X (a greater number of true positives and a reduced number of false negatives), but a lower specificity (fewer true negatives and more false positives), causing a larger number of nondiseased subjects to be incorrectly detected as positive (greater false-positive rate). The advantage, of course, is that fewer true positives will be missed. By contrast, setting the cutoff point higher (point B in the figure), the results will show greater specificity than with cutoff point X (a greater number of true negatives and a reduced number of false positives), but less sensitivity (fewer true positives and more false negatives), causing a lower number of diseased subjects to be detected (greater false-negative rate).

From a clinical perspective, the economic advantages of eliminating the cost of further screening for those who are not detected as being positive are far outweighed by the obvious disadvantage of the test's reduced sensitivity.

With respect to the regions of the curve pertaining to cutoff point X, choices F and E designate areas exclusively representing true positives and true negatives, respectively. Choice D designates an area with a mixture of true and false positives, and choice C an area with a mixture of true and false negatives.

Choice E also designates an area exclusively representing false positives when cutoff point A is employed, and choice F also designates an area exclusive for false negatives when using cutoff point B. For further explanation, refer to pages 18–19.

74–77. 74-(B), 75-(C), 76-(A), and 77-(D). The format chosen for the 2-by-2 tables in this review resembles the format chosen by most other references. However, it is important to note that the axes may easily be reversed for testing purposes.

Cells A, B, C, and D represent true positives, false positives, false negatives, and true negatives, respectively. For further explanation, refer to page 11.

78–84. 78-(E), 79-(B), 80-(F), 81-(D), 82-(A), 83-(C), and 84-(G).

Before ratio determinations can be recognized, a 2-by-2 table must be adequately organized. If 130 subjects were found to be positive for the test and 60 of them were false positives (cell B), the remaining 70 were true positives (cell A). If 30 of the 100 patients with the disease were false negatives (cell C), three of the four cells (A, B, and C) now include 160 (70 + 60 + 30) of the 200 volunteers in the study. The remainder of 40 volunteers represents the group who were true negatives (cell D). All the ratios given in the question can now be recognized. For further explanation, refer to page 16.

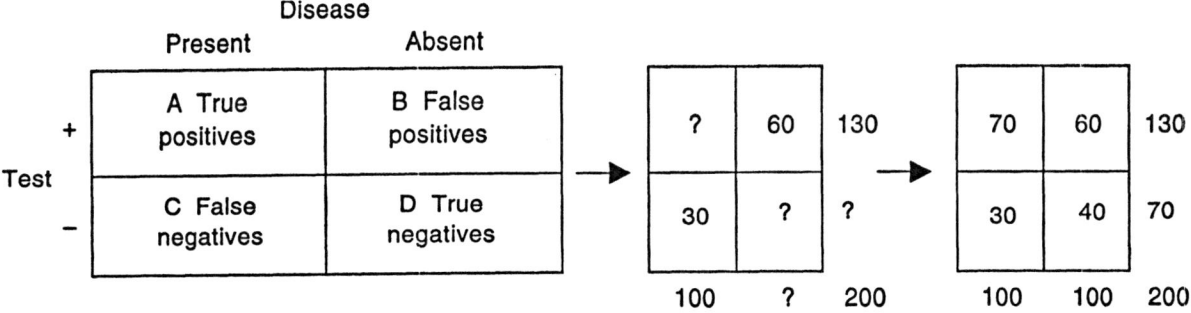

78. Sensitivity

$$= \frac{A}{A+C} = \frac{70}{70+30} = 0.70$$

79. Specificity

$$= \frac{D}{B+D} = \frac{40}{60+40} = 0.40$$

80. Positive predictive value

$$= \frac{A}{A+B} = \frac{70}{70+60} = 0.54$$

81. Negative predictive value

$$= \frac{D}{C+D} = \frac{40}{30+40} = 0.57$$

82. False-positive rate

$$= \frac{B}{B+D} = \frac{60}{60+40} = 0.60$$

83. False negative rate

$$= \frac{C}{A+C} = \frac{30}{70+30} = 0.30$$

84. Accuracy

$$= \frac{A+D}{A+B+C+D} = \frac{70+40}{70+60+30+40} = 0.55$$

85. **(D).** Diminished precision in a test is the result of random variability in test measurements. For further explanation, refer to page 11.

86. **(C).** The diminished accuracy of test measurements or sample classifications is usually the result of a nonrandom, systemic type of experimental error called bias. For further explanation, refer to page 12.

87-90. **87-(D), 88-(E), 89-(B), and 90-(C).**

87. A high positive correlation (scattergram D) represents a high positive relationship between two variables which can be seen by drawing a straight line through the central portion of the scattered data points. For each unit increase on the x axis, there is a nearly equivalent unit increase on the y axis.

88. A low positive correlation (scattergram E) represents a low positive relationship between two variables. Drawing a straight line through the central portion of the scattered data points reveals a very small unit increase on the y axis for each unit increase on the x axis.

89. A correlation coefficient of zero (scattergram B) represents no clear relationship between two variables. For each unit change on the x axis, the pattern of change on the y axis is indiscernible. A straight line drawn through the central portion of the scattered data points will lie horizontal to the x axis.

90. A high negative correlation (scattergram C) represents a high negative relationship between two variables that can be seen by drawing a straight line through the central portion of the scattered data points. For each unit increase on the x axis, there is a nearly equivalent unit decrease on the *y* axis. For further explanation, refer to pages 65–67.

91–94. 91-(E), 92-(A), 93-(F), and 94-(D). Validity is the appropriateness of a test—that is, its ability to measure what it was originally designed to measure. The validity of a test depends on the accuracy or trueness of its test measurements. Reliability is the ability of a test to be dependable—that is, fixed, durable, and well established. The reliability of a test depends on its precision (absence of random variability) to continually produce the same results (consistency) in many different groups of subjects (reproducibility). For further explanation, refer to pages 11–12.

95–99. 95-(E), 96-(C), 97-(B), 98-(B), and 99-(B). Accepting the null hypothesis when it is true and rejecting it when it is false are correct decisions. Rejecting the null hypothesis when it is true is a Type I (alpha) error, and accepting it when it is false is a Type II (beta) error. The beta error level is used to determine the probability of the test's ability to detect differences that actually exist (power of the test) by using the formula: power = 1 − beta. Because beta errors are sometimes the result of insignificant test results caused by inadequate sample sizes, the researcher may find that by increasing the size of the study sample a significant difference may in fact exist, the null hypothesis may be accurately rejected, and a Type II (beta) error avoided. For further explanation, refer to page 22.

100. (D). The primary purpose for the use of statistical methodologies in medical research is to determine the significance level of its findings. Statistical manipulations do not correct experimental errors, provide controls for variables, or estimate values for results that occur by chance. Statistical significance does not guarantee cause and effect relationships between variables.

APPENDIX A: Formula Summary

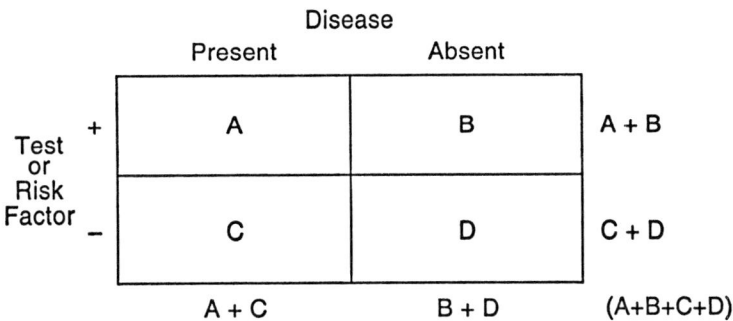

$$\text{Sensitivity} = \frac{\text{True positives}}{\text{True positives + False negatives}} = \frac{A}{A+C}$$

$$\text{Specificity} = \frac{\text{True negatives}}{\text{False positives + True negatives}} = \frac{D}{B+D}$$

$$\text{False-positive rate} = \frac{\text{False positives}}{\text{False positives + True negatives}} = \frac{B}{B+D}$$

$$\text{False-negative rate} = \frac{\text{False negatives}}{\text{True positives + False negatives}} = \frac{C}{A+C}$$

$$\text{Positive predictive value} = \frac{\text{True positives}}{\text{True positives + False positives}} = \frac{A}{A+B}$$

$$\text{Negative predictive value} = \frac{\text{True negatives}}{\text{True negatives + False negatives}} = \frac{D}{D+C}$$

$$\text{Accuracy of a test} = \frac{\text{True positives + True negatives}}{\text{All positives + All negatives}} = \frac{A+D}{(A+B+C+D)}$$

Absolute risk: *Risk group* = Incidence among exposed = $A/(A+B)$

Nonrisk group = Incidence among non-exposed = $C/(C+D)$
(Absolute risk is usually expressed as a percentage.)

$$\text{Relative risk} = \frac{\text{Incidence rate among risk group}}{\text{Incidence rate among nonrisk group}} = \frac{A/(A+B)}{C/(C+D)}$$

$$\text{Attributable risk} = \frac{\text{Incidence rate among risk group}}{} - \frac{\text{Incidence rate among non-risk group}}{}$$
$$= A/(A+B) - C/(C+D)$$

$$\text{Attributable risk percent} = \frac{\text{Attributable risk}}{\text{Absolute risk (risk group)}} \times 100$$

$$= \frac{A/(A+B) - C/(C+D)}{A/(A+B)} \times 100$$

$$\text{Odds ratio} = \frac{\text{Best estimate of relative risk}}{\text{in a retrospective (case control) study}} = \frac{A \times D}{B \times C}$$

$$\text{Incidence rate} = \frac{\text{Total number of new cases}}{\text{Total population at risk}}$$

$$\text{Prevalence rate} = \frac{\text{Number of existing cases}}{\text{Total midperiod population}}$$

$$\text{Prevalence} = \text{Incidence} \times \text{Duration of disease}$$

$$\text{Person-years} = \text{Number of persons} \times \text{Number of years followed}$$

$$\text{Case fatality rate} = \frac{\text{Number of deaths due to a disease}}{\text{Number of people with the same disease}} \times 10^x$$

$$\text{Proportionate mortality rate} = \frac{\text{Total number of deaths due to a disease}}{\text{Total number of deaths from all causes}} \times 10^x$$

$$\text{Cause-specific mortality rate} = \frac{\text{Number of deaths due to a disease}}{\text{Total midyear population}} \times 10^x$$

$$\text{Age-specific mortality rate} = \frac{\text{Number of deaths in that age group}}{\text{Population of same age group the same year}} \times 10^x$$

$$\text{Annual crude mortality rate} = \frac{\text{All deaths during a calendar year}}{\text{Total midyear population}} \times 10^x$$

$$\text{Infant mortality rate} = \frac{\text{Number of infant deaths} < 1 \text{ year old}}{\text{Total number of live births the same year}} \times 10^x$$

$$\text{Neonatal mortality rate} = \frac{\text{Number of neonatal deaths}}{\text{Total number of live births the same year}} \times 10^x$$

$$\text{Perinatal mortality rate} = \frac{\text{Number of perinatal deaths}}{\text{Total number of live births the same year}} \times 10^x$$

$$\text{Maternal mortality rate} = \frac{\text{Number of deaths from puerperal causes}}{\text{Total number of live births the same year}} \times 10^x$$

	Truth	
	H_0 True	H_0 False
Accept H_0	Correct	Type II error
Reject H_0	Type I error	Correct

Decision

Type I (alpha) Error: Rejecting H_0 when it is true
Type II (beta) Error: Accepting H_0 when it is false

APPENDIX B

Student's *t* Distribution Table

DEGREES OF	AREA IN 1 TAIL				
	0.05	0.025	0.01	0.005	0.0005
	AREA IN 2 TAILS				
FREEDOM	0.10	0.05	0.02	0.01	0.001
1	6.314	12.706	31.821	63.657	636.62
2	2.920	4.303	6.965	9.925	31.598
3	2.353	3.182	4.541	5.841	12.924
4	2.132	2.776	3.747	4.604	8.610
5	2.015	2.571	3.365	4.032	6.869
6	1.943	2.447	3.143	3.707	5.959
7	1.895	2.365	2.998	3.499	5.408
8	1.860	2.306	2.896	3.355	5.041
9	1.833	2.262	2.821	3.250	4.781
10	1.812	2.228	2.764	3.169	4.587
11	1.796	2.201	2.718	3.106	4.437
12	1.782	2.179	2.681	3.055	4.318
13	1.771	2.160	2.650	3.012	4.221
14	1.761	2.145	2.624	2.977	4.140
15	1.753	2.131	2.602	2.947	4.073
16	1.746	2.120	2.583	2.921	4.015
17	1.740	2.110	2.567	2.898	3.965
18	1.734	2.101	2.552	2.878	3.922
19	1.729	2.093	2.539	2.861	3.883
20	1.725	2.086	2.528	2.845	3.850
21	1.721	2.080	2.518	2.831	3.819
22	1.717	2.074	2.508	2.819	3.792
23	1.714	2.069	2.500	2.807	3.767
24	1.711	2.064	2.492	2.797	3.745
25	1.708	2.060	2.485	2.787	3.725
26	1.706	2.056	2.479	2.779	3.707
27	1.703	2.052	2.473	2.771	3.690
28	1.701	2.048	2.467	2.763	3.674
29	1.699	2.045	2.462	2.756	3.659
30	1.697	2.042	2.457	2.750	3.646
40	1.684	2.021	2.423	2.704	3.551
60	1.671	2.000	2.390	2.660	3.460
120	1.658	1.980	2.358	2.617	3.373
∞	1.645	1.960	2.326	2.576	3.291

[1] Adapted and reproduced, with permission, from Table 12 in Pearson ES, Hartley HO (editors): *Biometrika Tables for Statisticians*, 3rd ed. Vol 1. Cambridge University Press, 1966. Used with the kind permission of the Biometrika Trustees.

Appendix C

Chi-Square (χ^2) Distribution Table

DEGREES OF FREEDOM	AREA IN UPPER TAIL			
	0.10	0.05	0.01	0.001
1	2.706	3.841	6.635	10.828
2	4.605	5.991	9.210	13.816
3	6.251	7.815	11.345	16.266
4	7.779	9.488	13.277	18.467
5	9.236	11.071	15.086	20.515
6	10.645	12.592	16.812	22.458
7	12.017	14.067	18.475	24.322
8	13.362	15.507	20.090	26.125
9	14.684	16.919	21.666	27.877
10	15.987	18.307	23.209	29.588
11	17.275	19.675	24.725	31.264
12	18.549	21.026	26.217	32.909
13	19.812	22.362	27.688	34.528
14	21.064	23.685	29.141	36.123
15	22.307	24.996	30.578	37.697
16	23.542	26.296	32.000	39.252
17	24.769	27.587	33.409	40.790
18	25.989	28.869	34.805	42.312
19	27.204	30.144	36.191	43.820
20	28.412	31.410	37.566	45.315
21	29.615	32.671	38.932	46.797
22	30.813	33.924	40.289	48.268
23	32.007	35.173	41.638	49.728
24	33.196	36.415	42.980	51.179
25	34.382	37.653	44.314	52.620
26	35.563	38.885	45.642	54.052
27	36.741	40.113	46.963	55.476
28	37.916	41.337	48.278	56.892
29	39.088	42.557	49.588	58.302
30	40.256	43.773	50.892	59.703
40	51.805	55.759	63.691	73.402
50	63.167	67.505	76.154	86.661
60	74.397	79.082	88.379	99.607
70	85.527	90.531	100.425	112.317
80	96.578	101.879	112.329	124.839
90	107.565	113.145	124.116	137.208
100	118.498	124.342	135.807	149.449

[1] Adapted and reproduced, with permission, from Table 8 in Pearson ES, Hartley HO (editors): *Biometrika Tables for Statisticians,* 3rd ed. Vol 1. Cambridge University Press, 1966. Used with the kind permission of the Biometrika Trustees.

Appendix D

Commonly Used Statistical Symbols

α	alpha
α error	Type I error
β	beta
β error	Type II error
$1 - \beta$	1 minus beta (Power of a test)
x	sample mean
χ^2	chi-square value χ^2
μ	mu (mean of the standard population)
r	correlation coefficient
r^2	coefficient of determination
<	less than
>	greater than
$\leq$	less than or equal to
#	not equal to
Σ	sigma (the sum of the values)
d_f	degree of freedom
H_0	null hypothesis
H_A	alternate hypothesis
p	p value (significance level of a study or test)
s	S.D. (standard deviation of a sample)
s^2	variance (S.D. $\times$ S.D.)
n	sample size

INDEX

PRACTICE TEST ANSWER SHEET

1. A B C D E	51. A B C D E F
2. A B C D E	52. A B C D E F
3. A B C D E	53. A B C D E F
4. A B C D E	54. A B C D E F
5. A B C D E	55. A B C D E F
6. A B C D E	56. A B C D E
7. A B C D E	57. A B C D E
8. A B C D E	58. A B C D E
9. A B C D E	59. A B C D E
10. A B C D E	60. A B C D E F G H
11. A B C D E	61. A B C D E F G H
12. A B C D E	62. A B C D E F G H
13. A B C D E	63. A B C D E F G H
14. A B C D E	64. A B C D E F G H
15. A B C D E	65. A B C D E F G H
16. A B C D E	66. A B C D E F G H
17. A B C D E	67. A B C D E F G H
18. A B C D E	68. A B C D E F G H
19. A B C D E	69. A B C D E F G H
20. A B C D E	70. A B C D E F G H
21. A B C D E	71. A B C D E F G H
22. A B C D E	72. A B C D E F G H
23. A B C D E	73. A B C D E F G H
24. A B C D E	74. A B C D E
25. A B C D E	75. A B C D E
26. A B C D E	76. A B C D E
27. A B C D E	77. A B C D E
28. A B C D E	78. A B C D E F G
29. A B C D E	79. A B C D E F G
30. A B C D E	80. A B C D E F G
31. A B C D E	81. A B C D E F G
32. A B C D E	82. A B C D E F G
33. A B C D E	83. A B C D E F G
34. A B C D E	84. A B C D E F G
35. A B C D E	85. A B C D E
36. A B C D E	86. A B C D E
37. A B C D E	87. A B C D E
38. A B C D E	88. A B C D E
39. A B C D E	89. A B C D E
40. A B C D E	90. A B C D E
41. A B C D E	91. A B C D E F
42. A B C D E F	92. A B C D E F
43. A B C D E F	93. A B C D E F
44. A B C D E F	94. A B C D E F
45. A B C D E F	95. A B C D E F G
46. A B C D E	96. A B C D E F G
47. A B C D E	97. A B C D E F G
48. A B C D E	98. A B C D E F G
49. A B C D E	99. A B C D E F G
50. A B C D E F	100. A B C D E

Answers to Chapter Questions

1. D		31. F	
2. A		32. D	
3. B		33. E	
4. D		34. C	
5. A		35. E	
6. C		36. F	
7. D		37. A	
8. E		38. B	
9. B		39. C	
10. F		40. D	
11. G		41. C	
12. C		42. D	
13. C		43. D	
14. C		44. C	
15. A		45. D	
16. E		46. C	
17. D		47. D	
18. B		48. E	
19. C		49. C	
20. B		50. D	
21. D		51. D	
22. A		52. B	
23. A		53. E	
24. B		54. B	
25. A		55. C	
26. B		56. B	
27. B		57. C	
28. A		58. C	
29. A		59. B	
30. B		60. E	
		61. E	

Printed in the United States
113783LV00005B/112/A

9 780838 502440